PMP® Practice Makes Perfect

Over 1,000 PMP® Practice Questions and Answers

John A. Estrella
Charles Duncan
Sami Zahran
James L. Haner
Rubin Jen

John Wiley & Sons, Inc.

Senior Acquisitions Editor: Jeff Kellum
Development Editor: Jennifer Leland
Technical Editors: Vanina Mangano and Dan Gindin
Production Editor: Dassi Zeidel
Copy Editor: Liz Welch
Editorial Manager: Pete Gaughan
Production Manager: Tim Tate
Vice President and Executive Group Publisher: Richard Swadley
Vice President and Publisher: Neil Edde
Book Designer: Judy Fung and Bill Gibson
Compositor: Craig Johnson, Happenstance Type-O-Rama
Proofreader: Sara Wilson
Indexer: Ted Laux
Project Coordinator, Cover: Katherine Crocker
Cover Designer: Ryan Sneed

Published simultaneously in Canada

ISBN: 978-1-118-16976-6
ISBN: 978-1-118-22658-2 (ebk.)
ISBN: 978-1-118-23971-1 (ebk.)
ISBN: 978-1-118-26441-6 (ebk.)

Library of Congress Control Number: 2011943912

10 9 8 7 6 5 4 3 2

Dear Reader,

Thank you for choosing *PMP® Practice Makes Perfect: Over 1,000 PMP® Practice Questions and Answers*. This book is part of a family of premium-quality Sybex books, all of which are written by outstanding authors who combine practical experience with a gift for teaching.

Sybex was founded in 1976. More than 30 years later, we're still committed to producing consistently exceptional books. With each of our titles, we're working hard to set a new standard for the industry. From the paper we print on, to the authors we work with, our goal is to bring you the best books available.

I hope you see all that reflected in these pages. I'd be very interested to hear your comments and get your feedback on how we're doing. Feel free to let me know what you think about this or any other Sybex book by sending me an email at `nedde@wiley.com`. If you think you've found a technical error in this book, please visit `http://sybex.custhelp.com`. Customer feedback is critical to our efforts at Sybex.

Best regards,

Neil Edde
Vice President and Publisher
Sybex, an Imprint of Wiley

We would like to dedicate this book to the founders and volunteers of the Project Management Institute (PMI®).

—John, Charles, Sami, James, and Rubin

Acknowledgments

We would like to thank Luisito Pangilinan, A. J. Sobzak, Roserene Balana, Norma Nieves, Raymond Chung, and Barry Mascoe for their assistance with this book. Jeff Kellum, Jennifer Leland, Dassi Zeidel, Liz Welch, Pete Gaughan, Tim Tate, Richard Swadley, Neil Edde, Judy Fung, Bill Gibson, Craig Johnson, Sara Wilson, Ted Laux, Katherine Crocker, Ryan Sneed, and the rest of the team at Sybex who made this book possible—we cannot thank you enough! Our technical editors extraordinaire, Vanina Mangano and Dan Gindin, deserve a special mention for diligently reviewing all the questions for technical accuracy and providing valuable suggestions to ensure that the questions, options, and explanations are clear for the reader.

About the Authors

John A. Estrella, PhD, CMC, PMP® is the president of Agilitek Corporation, a management consulting firm that specializes in project management, business analysis, and software testing. As an international consultant, author, and speaker, John helps senior managers and executives in addressing issues associated with complex, large-scale, and risky projects.

Charles Duncan, PhD, PMP®, APM Practitioner Qualification manages a consulting group that specializes in software contract development, technical project troubleshooting, and management training. Charlie has considerable experience with the management of start-up and high-technology organizations and project management at ICL/Fujitsu, Hewlett-Packard R&D Labs, and Wang Laboratories.

Sami Zahran, PhD, PMP® has over three decades of experience in the software industry assuming senior positions with large organizations including ICL, the United Nations, DEC, and currently IBM. Sami regularly teaches courses on his areas of expertise and frequently speaks at numerous international conferences and workshops.

James L. Haner, PgMP, PMP® is the head of Ultimate Business Resources (UBR), a firm that offers business and project management consulting services to Fortune 500 companies in the United States, Europe, Africa, and China. James has more than 30 years of experience in business and IT.

Rubin Jen, BASc, P. Eng, PMP® has been in project management for over 15 years, spanning aerospace/engineering, telecom, information technology, outsourcing, and consulting industries with companies such as Bombardier Aerospace, Celestica, Organic Online, OnX Enterprise Solutions, and Accenture. Rubin is a speaker, writer, and instructor of project management.

Contents

Introduction

The Project Management Institute (PMI®) is the leader and most widely recognized organization in terms of promoting project management best practices. PMI® strives to maintain and endorse standards and ethics in this field and offers publications, training, seminars, chapters, special interest groups, and colleges to further the project management discipline.

PMI® was founded in 1969 and first started offering the Project Management Professional (PMP®) certification exam in 1984. PMI® is accredited as an American National Standards Institute (ANSI) standards developer and also has the distinction of being the first organization to have its certification program attain International Organization for Standardization (ISO) 9001 recognition.

PMI® boasts a worldwide membership of more than 265,000, with members from 170 different countries. Local PMI® chapters meet regularly and allow project managers to exchange information and learn about new tools and techniques of project management or new ways to use established techniques. We encourage you to join a local chapter and get to know other professionals in your field.

Why Become PMP® Certified?

The following benefits are associated with becoming PMP® certified:

- It demonstrates proof of professional achievement.
- It increases your marketability.
- It provides greater opportunity for advancement in your field.
- It raises customer confidence in you and in your company's services.

Demonstrates Proof of Professional Achievement

PMP® certification is a rigorous process that documents your achievements in the field of project management. The exam tests your knowledge of the disciplined approaches, methodologies, and project management practices as described in *A Guide to the Project Management Body of Knowledge (PMBOK® Guide)*, 4th Edition.

You are required to have several years of experience in project management before sitting for the exam, as well as 35 hours of formal project management education. Your certification assures employers and customers that you are well grounded in project management practices and disciplines. It shows that you have the hands-on experience and a mastery of the processes and disciplines to manage projects effectively and motivate teams to produce successful results.

Increases Your Marketability

Many industries are realizing the importance of project management and its role in the organization. They are also seeing that simply proclaiming a head technician to be a *project manager* does not make it so. Project management, just like engineering, information technology, and a host of other trades, has its own specific qualifications and skills. Certification tells potential employers that you have the skills, experience, and knowledge to drive successful projects and ultimately improve the company's bottom line.

A certification will always make you stand out above the competition. If you're certified and you're competing against a project manager without a certification, chances are you'll come out as the top pick. Hiring managers, all other things being equal, will usually opt for the candidate who has certification over the candidate who doesn't have it. Certification tells potential employers that you have gone the extra mile. You've spent time studying techniques and methods as well as employing them in practice. It shows dedication to your own professional growth and enhancement and adherence to advancing professional standards.

Provides Opportunity for Advancement

PMP® certification displays your willingness to pursue growth in your professional career and shows that you're not afraid of a little hard work to get what you want. Potential employers will interpret your pursuit of this certification as a high-energy, success-driven, can-do attitude on your part. They'll see that you're likely to display these same characteristics on the job, which will help make the company successful. Your certification displays a success-oriented, motivated attitude that will open up opportunities for future career advancements in your current field as well as in new areas you might want to explore.

Raises Customer Confidence

Just as the PMP® certification assures employers that you've got the background and experience to handle project management, it assures customers that they have a competent, experienced project manager at the helm. Certification will help your organization sell customers on your ability to manage their projects. Customers, like potential employers, want the reassurance that those working for them have the knowledge and skills necessary to carry out the duties of the position and that professionalism and personal integrity are of utmost importance. Individuals who hold these ideals will translate their ethics and professionalism to their work. This enhances the trust customers will have in you, which in turn will give you the ability to influence them on important project issues.

How to Become PMP® Certified

You need to fulfill several requirements in order to sit for the PMP® exam. The PMI® has detailed the certification process quite extensively at its website. Go to `www.pmi.org`, and click the Certifications tab to get the latest information on certification procedures and requirements.

As of this writing, you are required to fill out an application to sit for the PMP® exam. You can submit this application online at the PMI®'s website. You also need to document 35 hours of formal project management education. This might include college classes, seminars, workshops, and training sessions. Be prepared to list the class titles, location, date, and content.

In addition to filling out the application and documenting your formal project management training, you need to meet the criteria in one of the following two categories to sit for the exam:

- Category 1 is for those who have a baccalaureate degree. You'll need to provide proof, via transcripts, of your degree with your application. In addition, you'll need to complete verification forms—found at the PMI® website—that show 4,500 hours of project management experience that spans a minimum of three years and no more than six years.
- Category 2 is for those who do not have a baccalaureate degree but do hold a high school diploma or equivalent. You'll need to complete verification forms documenting 7,500 hours of project management experience that spans a minimum of five years and no more than eight years.

As of this writing, the exam fee is a little over $400 for PMI® members in good standing and less than $600 for non-PMI® members. Testing is conducted at Thomson Prometric centers. You can find a center near you on the PMI® website. You have six months from the time PMI® receives and approves your completed application to take the exam. You'll need to bring a form of identification such as a driver's license with you to the Thomson Prometric center on the test day. You will not be allowed to take anything with you into the testing center. You will be given a calculator, pencils, and scrap paper. You will turn in all scrap paper, including the notes and squiggles you've jotted during the test, to the center upon completion of the exam.

The exam is scored immediately, so you will know whether you've passed at the conclusion of the test. You're given four hours to complete the exam, which consists of 200 randomly generated questions. Only 175 of the 200 questions are scored. A passing score requires you to answer 106 of the 175 questions correctly. Twenty-five of the 200 questions are "pretest" questions that will appear randomly throughout the exam. These 25 questions are used by PMI® to determine statistical information and to determine whether they can or should be used on future exams. The questions on the exam cover the following process groups and areas:

- Initiating
- Planning
- Executing
- Monitoring and Controlling
- Closing
- Professional Responsibility

Questions pertaining to professional responsibility on the exam will be intermixed with all the other process groups previously listed. You won't see a section or set of questions devoted solely to professional responsibility, but you will need to understand all the concepts in this area.

The following table shows the breakdown of questions by process groups of the actual exam.

BREAKDOWN OF PMP® QUESTIONS

Process groups	Percent of questions	Number of questions
1. Initiating the project	13%	26
2. Planning the project	24%	48
3. Executing the project	30%	60
4. Monitoring and controlling the project	25%	50
5. Closing the project	8%	16
Total	100%	200

If you are unsure how to answer a question, use the process of elimination. Eliminate one choice that you know is incorrect. This simple trick will increase your likelihood of guessing the correct answer from 25% to 33%. For the remaining choices, determine which one seems *correct* but is not appropriate or relevant for the question. After you do this, you will be left with two choices, thus giving you a 50% chance of answering the question correctly. If the two remaining choices appear to be both valid or correct, make an educated guess and pick the answer that is *more correct* than the other. In contrast, you can also select an answer that is *least incorrect* compared to the other remaining choice.

As you go through each question, make sure to pace yourself properly. If you find that you are spending too much time on a question, just mark it and proceed to the next question. You can always go back later. Lastly, before you start the exam, write down all formulas, acronyms, and other terms on a blank piece of paper that will be provided to you. It is best to do this when your mind is still fresh. This technique will enable you to easily recall important information later on.

After you receive your certification, you'll be required to earn 60 professional development units (PDUs) every three years to maintain certification. Approximately one hour of structured learning translates to one PDU. The PMI® website details what activities constitute a PDU, how many PDUs each activity earns, and how to register your PDUs with PMI® to maintain your certification. As an example, attendance at a local chapter meeting earns one PDU.

How to Use This Book

This book was designed to help you prepare, in a time-efficient manner, for PMI®'s PMP® exam. The distilled format of this book will provide you with the fundamental knowledge in project management as an aid to passing the exam with the minimum amount of effort.

We wrote this book with four categories of readers in mind: planners, crammers, refreshers, and teachers:

Planners Planners are very deliberate in their approach to taking the exam. They thoroughly research the topic and allocate sufficient time to prepare for the exam. These types of readers can quickly dive right into Chapter 3, "Practice Test A," Chapter 5, "Practice Test B," Chapter 7, "Practice Test C," and Chapter 9, "Practice Test D," to try the four sets of sample exam questions. If you feel confident about taking the actual exam after answering the practice questions, you may skip the solutions in Chapter 4, "Answers to Practice Test A," Chapter 6, "Answers to Practice Test B," Chapter 8, "Answers to Practice Test C," and Chapter 10 "Answers to Practice Test D."

Crammers Crammers should make sure to carefully read the information in this Introduction before attempting the sample exam questions in Chapter 3, Chapter 5, Chapter 7, and Chapter 9. If you did not score well on the practice questions, carefully read the solutions in Chapter 4, Chapter 6, Chapter 8, and Chapter 10.

Refreshers If you previously failed the exam or have taken an exam preparation course but would like to further boost your confidence, we consider you to be part of the refreshers group. Try the sample exam questions to determine your areas of strength and weakness. Read the corresponding sections in the *PMBOK® Guide* for your weak areas and focus on the same topics in the solution chapters.

Teachers Teachers may use the book as a valuable course supplement. They can measure students' understanding of each knowledge area by focusing on specific questions from Chapter 4, Chapter 6, Chapter 8, and Chapter 10.

Note that Chapter 4, Chapter 6, Chapter 8, and Chapter 10 include an answer key, so you can quickly check your answers against the correct answers. In addition, we have provided a PDF of an answer bubble sheet on the book's website so you don't have to mark the answers in your book. This PDF can be downloaded at `www.sybex.com/go/pmppractice`.

Chapter 1, "Quick Review Questions," and Chapter 2, "Quick Review Solutions," are meant to be a quick review before you try the practice questions. Make sure that you can confidently answer all the questions in Chapter 1 and that you have reviewed the correct answers in Chapter 2 before moving on to the remaining chapters.

Additional Resources

To ensure accurate adherence to the exam requirements, this book relied heavily on the contents of *PMBOK® Guide,* 4th Edition and PMI®'s *Code of Ethics and Professional Conduct*, which is included in the *Project Management Professional (PMP®) Handbook*. Each question, in addition to a detailed explanation, includes a reference to the relevant resource.

We also strictly followed the *Project Management Professional (PMP®) Examination Content Outline* and ensured alignment to the Role Delineation Study.

We used the following references in the book:

Heldman, K. *PMP®: Project Management Professional Exam Study Guide* (6th ed.). Sybex, 2011.

Project Management Institute, Inc. (Pennsylvania, 2008). *A guide to the project management body of knowledge (PMBOK® Guide)* (4th ed.).

Project Management Institute, Inc. (Pennsylvania, 2010). *Project management professional (PMP®) examination content outline.*

Project Management Institute, Inc. (Pennsylvania, 2011). *Project management professional (PMP®) handbook.*

The *PMP® Handbook* and *PMP® Examination Content Outline* can be downloaded from PMI®'s website at `www.pmi.org`.

How to Contact the Authors

If you have any questions about this book, please feel free to contact the lead author, John A. Estrella, directly at `jestrella@agilitek.com`.

Chapter 1

Quick Review Questions

The following questions are categorized by the knowledge areas of *A Guide to the Project Management Body of Knowledge, 4th Edition (PMBOK® Guide)*. Take note that some questions came from the Appendix. There are also questions from the Project Management Institute's (PMI) *Code of Ethics and Professional Conduct*. Before you sit for the Project Management Professional (PMP)® exam, you should be familiar with these publications and what is covered in them. The following questions are designed to test your familiarity with them, as well as your understanding and knowledge of general project management terms and concepts.

A Guide to the Project Management Body of Knowledge, 4th Edition

General Knowledge of *PMBOK Guide®*

1. At a minimum, you need to be familiar with at least three PMI publications before taking the PMP exam. List them in the following spaces.

   A. ______________________________

   B. ______________________________

   C. ______________________________

2. List the five process groups in the order that they appear in the *PMBOK® Guide*.

   A. ______________________________

   B. ______________________________

   C. ______________________________

   D. ______________________________

   E. ______________________________

3. List the nine Knowledge Areas in the order they appear in the *PMBOK® Guide*.

A. ______________________________

B. ______________________________

C. ______________________________

D. ______________________________

E. ______________________________

F. ______________________________

G. ______________________________

H. ______________________________

I. ______________________________

4. List the three key components of a process as outlined in the *PMBOK® Guide*.

A. ______________________________

B. ______________________________

C. ______________________________

5. Organizational structure is an enterprise environmental factor that influences a project. List the six common organizational structures.

A. ______________________________

B. ______________________________

C. ______________________________

D. ______________________________

E. ______________________________

F. ______________________________

6. List the two processes that should be performed when starting a project.

A. ______________________________

B. ______________________________

7. List the two processes that should be performed when closing a project.

A. ______________________________

B. ______________________________

8. The *PMBOK® Guide* consistently uses the verb-noun format to describe each process. If you carefully examine the verb portion of the process name, it should give you an indication that a given process belongs to a certain process group. The noun portion of the process name should also give you a hint of the expected deliverable for that process. For example, the Develop Project Management Plan process indicates that you need to "develop" something after you have started your project—with the "Project Management Plan" as the expected deliverable. List the 10 common verbs that are used to name the processes within the Planning process group.

9. List the seven common verbs that are used to name the processes within the Initiating process group.

10. List the six common verbs that are used to name the processes within the Monitoring and Controlling process group.

Chapter 4: Project Integration Management

11. List at least 10 subsidiary plans of the project management plan.

Chapter 5: Project Scope Management

12. List the eight tools and techniques that a project manager can use when collecting requirements.

13. List three terms that relate to the work breakdown structure (WBS).

A. ______________________________

B. ______________________________

C. ______________________________

Chapter 6: Project Time Management

14. What are the four types of activity dependencies or logical relationships?

A. ______________________________

B. ______________________________

C. ______________________________

D. ______________________________

15. List five estimating tools and techniques.

A. ______________________________

B. ______________________________

C. ______________________________

D. ______________________________

E. ______________________________

16. What is the PERT formula for three-point estimating?

____

Chapter 7: Project Cost Management

Define the following Earned Value Management (EVM) acronyms.

17. What is PV? ______________________________

18. What is EV? ______________________________

19. What is AC? ______________________________

20. What is SV? ______________________________

21. What is CV? ______________________________

22. What is SPI? ______________________________

23. What is CPI? ______________________________

24. What is EAC? ______________________________

25. What is ETC? ______________________________

26. What is TCPI? ______________________________

27. List the two EVM variance formulas.

A. ______________________

B. ______________________

28. List the two EVM performance index formulas.

A. ______________________

B. ______________________

29. List the four EVM forecasting formulas for EAC.

A. ______________________

B. ______________________

C. ______________________

D. ______________________

30. List the two to-complete performance index formulas.

A. ______________________

B. ______________________

Chapter 8: Project Quality Management

Define the following quality-specific acronyms.

31. What is PDCA? ______________________

32. What is TQM? ______________________

33. What is OPM3? ______________________

34. What is CMMI? ______________________

35. What is COQ? ______________________

36. Name the two costs of quality.

A. ______________________

B. ______________________

37. What are the two categories of the cost of quality for avoiding failures?

A. ______________________

B. ______________________

38. What are the two categories of the cost of quality where money must be spent because of failures?

A. ______________________

B. ______________________

39. List Ishikawa's seven basic tools of quality.

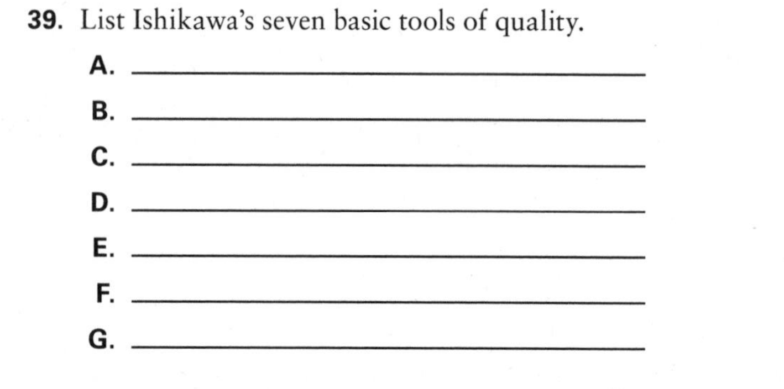

A. ______________________________

B. ______________________________

C. ______________________________

D. ______________________________

E. ______________________________

F. ______________________________

G. ______________________________

Chapter 9: Project Human Resource Management

40. List in order the five stages of developing a team.

A. ______________________________

B. ______________________________

C. ______________________________

D. ______________________________

E. ______________________________

41. What are the six general techniques for avoiding conflict?

A. ______________________________

B. ______________________________

C. ______________________________

D. ______________________________

F. ______________________________

G. ______________________________

Chapter 10: Project Communications Management

42. What are the five key components of a communication model?

A. ______________________________

B. ______________________________

C. ______________________________

D. ______________________________

E. ______________________________

43. List the three broad classifications of communication methods.

A. ______________________________

B. ______________________________

C. ______________________________

Chapter 11: Project Risk Management

44. List four strategies for negative risks or threats.

A. ______________________________

B. ______________________________

C. ______________________________

D. ______________________________

45. List four strategies for positive risks or opportunities.

A. ______________________________

B. ______________________________

C. ______________________________

D. ______________________________

Chapter 12: Project Procurement Management

46. What are the three major types of contracts?

A. ______________________________

B. ______________________________

C. ______________________________

There are six potential variations of the major types of contracts.

47. What is FFP? ______________________________

48. What is FPIF? ______________________________

49. What is FP-EPA? ______________________________

50. What is CPFF? ______________________________

51. What is CPIF? ______________________________

52. What is CPAF? ______________________________

Appendix G: Interpersonal Skills

53. List at least five interpersonal skills.

A. ______________________

B. ______________________

C. ______________________

D. ______________________

E. ______________________

54. List at least three team member communication styles.

A. ______________________

B. ______________________

C. ______________________

55. What are the two categories of listening techniques?

A. ______________________

B. ______________________

56. What are the four basic decision styles?

A. ______________________

B. ______________________

C. ______________________

D. ______________________

57. What are the four major factors that affect the decision styles?

A. ______________________

B. ______________________

C. ______________________

D. ______________________

Define the following acronyms.

58. What is CCB? ______________________

59. What is CPM? ______________________

60. What is EMV? ______________________

61. What is FMEA? ______________________

62. What is IFB? ______________________

63. What is LOE? ______________________

64. What is OBS? ______________________

65. What is PDM? ______________________

66. What is QA? ______________________

67. What is QC? ______________________

68. What is RACI? ______________________

69. What is RAM? ______________________

70. What is RBS? ______________________

71. What is RFI? ______________________

72. What is RFP? ______________________

73. What is RFQ? ______________________

74. What is SOW? ______________________

75. What is SWOT? ______________________

PMI's *Code of Ethics and Professional Conduct*

76. The PMI's *Code of Ethics and Professional Conduct* applies to which individuals?

    A. ______________________

    B. ______________________

    C. ______________________

    D. ______________________

77. What two types of responsibility, fairness, and honesty standards were mentioned in the *Code of Ethics and Professional Conduct*?

    A. ______________________

    B. ______________________

Key Concepts and Processes

Many of these questions discuss topics and concepts presented in Sybex's *PMP: Project Management Professional Exam Study Guide, 6th Edition* (ISBN: 978-1118083215).

This section presents questions based on key project management concepts (e.g., projects, programs, portfolios, project management skills, etc.) and various processes from initiating to closing projects.

Projects

78. _______________ are temporary initiatives with defined start and end dates.

79. The concept of incrementally and continually refining the characteristics of a product, service, or result as the project progresses is known as _______________.

80. Unlike projects, _______________ are ongoing and repetitive.

81. _______________ are individuals or groups with a vested interest in the outcome of the project.

82. The _______________ is usually an executive within the organization who has authority over the project, from allocating resources to making key decisions.

83. The _______________ is responsible for applying the tools and techniques of project management, and managing the associated resources to ensure that the project objectives will be met.

84. _______________ are groups of related projects.

85. _______________ include projects, programs, and other portfolios.

86. The _______________ serves as a central oversight in the management of projects and programs within an organization.

87. What is a key disadvantage for a project manager when working in a functional organization? ______________________________________

88. _______________ are considered the opposite of functional organizations.

89. What are the three common types of matrix structures?

A. _______________

B. _______________

C. _______________

Creating the Project Charter

Fill in the blanks or answer the questions below.

90. Who is not involved in project selection: project manager, project sponsor, customer, or subject matter experts? ______________

91. ______________ are also used to evaluate and choose between alternative ways of performing the project.

92. What is the discount rate when NPV equals zero? ______________

93. Which project should be chosen: low IRR or high IRR?

____

94. The ______________ process produces the project charter.

95. Corporate procedures, policies, standards, and templates are elements of

______________.

96. Who should be the author of the project charter?

____

97. Projects that have an NPV ______________ zero have a higher likelihood of being approved.

98. ______________ are external factors that can positively or negatively impact the outcome of the project.

99. Name at least two needs or demands that necessitate the initiation of a project.

A. ______________

B. ______________

C. ______________

D. ______________

E. ______________

F. ______________

G. ______________

Developing the Project Scope Statement

Determine if each of the following statements is true or false. If the statement is false, provide the correct answer to the underlined term(s) to make the statement true.

100. The <u>project charter</u> defines how the project will be executed, monitored and controlled, and closed

____

101. Requirements must be documented, analyzed, and quantified in enough detail that they can be measured once the work of the project begins.

____

102. Focus groups comprise cross-functional stakeholders.

____

103. The requirements traceability matrix enables the organization to link each requirement to business and project objective.

____

104. The project scope management plan documents how the project team will define the project scope, develop the work breakdown structure, and control the changes.

____

105. The Create WBS process enables the project team to subdivide the project into smaller and more manageable components.

____

106. Rolling wave planning is a process of elaborating deliverables, project phases, or subprojects in the WBS.

____

107. The scope baseline includes the scope statement, the WBS, and the project schedule.

____

108. The scope management plan is a subsidiary plan of the project charter.

____

109. The lowest level of any WBS is called a work package.

____

Creating the Project Schedule

Match the terms below to their correct descriptions. Write the letter of the correct description in the space provided.

110. Analogous estimating ______

111. Define Activities process ______

112. Discretionary dependency ______

113. Finish-to-start ______

114. Lag ______

115. Lead ______

116. Monte Carlo ______

117. One standard deviation ______

118. PDM ______

119. Start-to-finish ______

Descriptions

A. Creates arbitrary total float values

B. Decomposes the work packages into schedule activities

C. Delays successor activities

D. Gives a probability of about 68 percent

E. Most commonly used dependency in the PDM method

F. Rarely used dependency in the PDM method

G. Shows probability of all the possible project completion dates

H. Speeds up successor activities

I. Uses expert judgment and historical information

J. Uses only one time estimate to determine duration

Developing the Project Budget

Fill in the blanks or answer the questions below.

120. The ______________________ is established using the WBS and its associated control accounts.

121. The control account also has a unique identifier that's linked to the organization's accounting system, sometimes known as a ______________________.

122. ______________________ should be completed early in the project to help estimate the costs.

123. The following three additional tools can be used to predict the potential financial performance of the project's product, service, or result: ______________________.

124. A project's cost performance baseline normally has a/an ______________________ shape.

125. ______________________ entails providing the information in the correct format to the right audience in a timely manner.

126. What is the communication channel formula?

____

127. The lines that connect the communication nodes are also known as ______________________________.

128. The ______________________________ addresses the communication needs of the project stakeholders.

129. What is a PMB? ______________________________

Risk Planning

Determine if each statement below is true or false. If the statement is false, provide the correct answer to the underlined term(s) to make the statement true.

130. <u>Not all</u> risks are bad.

____

131. The <u>Plan Risk Management</u> process serves as the foundation for subsequent risk processes.

____

132. RBS is an acronym for <u>requirements budgeting system</u>.

____

133. Catastrophic risks are also known as <u>residual risks</u>.

____

134. <u>Information gathering</u> is a type of brainstorming.

____

135. <u>The fishbone diagram</u> is also known as the Ishikawa diagram.

____

136. <u>Probability</u> is the likelihood that an event will occur.

____

137. Sensitivity analysis data can be displayed using a <u>sensitivity diagram</u>.

____

138. The sum of the probabilities for each node in a decision tree is <u>1</u>.

____

139. The following strategies can be used to deal with negative risks or threats: avoid, transfer, mitigate, and share.

____

Planning Project Resources

Match the terms below to their correct descriptions. Write the letter of the correct description in the space provided.

140. Collective bargaining agreements ______

141. Firm fixed-price ______

142. FP-EPA economic adjustment ______

143. Make-or-buy analysis ______

144. Organization breakdown structure ______

145. Project calendars ______

146. Regulation ______

147. Source selection criteria ______

148. Teaming agreement ______

149. Time and materials contracts ______

Descriptions

A. Crosses between fixed-price and cost-reimbursable contracts

B. Defines the contractual agreements between multiple parties

C. Defines the organization's contractual obligations to employees

D. Facilitates the decision-making process whether to build or purchase products and services that are needed in the project

E. Helps organizations choose a vendor from among the proposals received

F. Mandates linking it to a known financial index

G. Places the bulk of the risks on the seller

H. Requires compliance and are almost always imposed by governments or institutions

I. Shows the departments, work units, etc. within an organization

J. Shows the working and nonworking days in the project

Developing the Project Team

Fill in the blanks or answer the questions below.

150. Which comes first: validated defect repair or defect repair?

____

151. Which comes first: norming or storming?

____

152. Which comes first: storming or forming?

____

153. To achieve co-location, the project manager may set aside a common meeting area, which is sometimes called the ______________________________.

154. Motivation can be ______________________________ or ______________________________.

155. Which one is higher in the hierarchy of needs: social needs or basic physical needs?

____

156. Which one is higher in the hierarchy of needs: safety and security needs, or self-actualization?

____

157. ______________________________ Theory: positive outcomes drive motivation.

____

158. Which type of power should be used as a last resort: legitimate or punishment?

____

159. What are the most important skills that a project manager should possess?

____

Coordinating Procurements and Sharing Information

Fill in the blanks or answer the questions below.

160. Which proposal evaluation technique assigns numerical weights to evaluation criteria: weighting systems or screening systems?

____

161. Which proposal evaluation technique uses predefined performance criteria: weighting systems or screening systems?

____

162. Which proposal evaluation technique relies on information about the sellers: screening systems or seller rating systems?

____

163. Independent estimates, sometimes conducted by the procurement department to compare the proposal costs against the vendor costs, are also known as

____________________________________.

164. What tactic is used during contract negotiation when one party tries to discuss an issue that is no longer relevant?

____

165. Which contract life-cycle stage comes first: solicitation or requisition?

____

166. Which contract life-cycle stage comes first: requirement or award?

____

167. ____________________________________ are independent reviews performed by trained auditors or third-party reviewers.

168. The ____________________________________ helps promote communication with stakeholders.

169. Contested changes can take form of ____________________________________, ____________________________________, or ____________________________________.

Measuring and Controlling Project Performance

Fill in the blanks or answer the questions below.

170. A ____________________________________ is a subsystem of the project management system.

171. To get paid, vendors issue ____________________________________.

172. When a seller submits an invoice, the ____________________________________ is used to issue payments.

173. What is ADR? ____________________________________

174. ____________________________________ use historical data to predict future performance.

175. ____________________________________ is a forecasting method that uses opinions, intuition, and estimates to predict future outcomes.

176. Who is usually the expediter of project status meeting?

____

177. What do you need to do to introduce modifications to the project?

____

178. Is the change control system a subset of the configuration management system?

____

179. Which one is the preferred approach for capturing change requests: written or verbal?

____

Controlling Work Results

Determine if each statement below is true or false. If the statement is false, provide the correct answer to the underlined term(s) to make the statement true.

180. Planned value is the same as <u>budgeted cost of work scheduled</u>.

____

181. Actual cost is the same as <u>actual cost of work performed</u>.

____

182. Earned value is the same as <u>actual cost of work performed</u>.

____

183. Given an EV of 350 and AC of 300, the CV is <u>650</u>.

____

184. A CPI greater than <u>1</u> is good.

____

185. EAC = <u>AC + BAC – EV</u>.

____

186. <u>VAC</u> = BAC – EAC.

____

187. <u>CV</u> = EV – PV.

____

188. A <u>control chart</u> has lower and upper control limits.

____

189. Ishikawa is credited with the 80/20 Rule.

____

190. A scatter diagram's two variables are dependent and independent variables.

____

Closing the Project and Applying Professional Responsibility

Match the terms below with their correct descriptions. Write the letter of the correct description in the space provided.

191. Addition, starvation, integration, and extinction ______

192. Aspirational and mandatory ______

193. Closed procurements ______

194. Lessons Learned ______

195. Organization process updates ______

196. PMI's *Code of Ethics and Professional Conduct* ______

197. Procurement audit ______

198. Procurement early termination ______

199. Project management plan ______

200. Responsibility, respect, fairness, and honesty ______

Descriptions

A. Applicable to both buyer and vendor

B. Areas of professional responsibility

C. By agreement, via default or for cause

D. Captures what went well in the project and what should be avoided for the next project

E. Ethical standards for PMPs

F. Formal acceptance and closure of the procurement

G. Input of Close Procurements process

H. Output of the Close Project or Phase process

I. Standards of professional responsibility

J. Types of project endings

Chapter

2

Quick Review Solutions

The following questions are categorized by the knowledge areas of *A Guide to the Project Management Body of Knowledge, 4th Edition (PMBOK® Guide)*. Take note that some questions came from the Appendix. There are also questions from the Project Management Institute's (PMI) *Code of Ethics and Professional Conduct*. Before you sit for the Project Management Professional (PMP)® exam, you should be familiar with these publications and what is covered in them. The following questions are designed to test your familiarity with them, as well as your understanding and knowledge of general project management terms and concepts.

A Guide to the Project Management Body of Knowledge, 4th Edition Answers

The page numbers refer to the page number of the corresponding text in the *PMBOK Guide®, 4th Edition.*

General Knowledge of *PMBOK Guide®*

1. At a minimum, you need to be familiar with at least three PMI publications before taking the PMP exam. List them in the following spaces.
   - **A.** *Project Management Professional (PMP) Handbook* (includes the PMI *Code of Ethics and Professional Conduct*)
   - **B.** *A Guide to the Project Management Body of Knowledge (PMBOK® Guide)—4th Edition*
   - **C.** *Project Management Professional (PMP) Examination Content Outline*
2. List the five process groups in the order that they appear in the *PMBOK® Guide* (page 43).
   - **A.** Initiating
   - **B.** Planning
   - **C.** Executing
   - **D.** Monitoring and Controlling
   - **E.** Closing

3. List the nine Knowledge Areas in the order they appear in the *PMBOK® Guide* (page 43).

   A. Project Integration Management

   B. Project Scope Management

   C. Project Time Management

   D. Project Cost Management

   E. Project Quality Management

   F. Project Human Resource Management

   G. Project Communications Management

   H. Project Risk Management

   I. Project Procurement Management

4. List the three key components of a process as outlined in the *PMBOK® Guide* (page 3).

   A. Inputs

   B. Tools and techniques

   C. Outputs

5. Organizational structure is an enterprise environmental factor that influences a project. List the six common organizational structures (pages 28–31).

   A. Functional organization

   B. Weak matrix organization

   C. Balanced matrix organization

   D. Strong matrix organization

   E. Projectized organization

   F. Composite organization

6. List the two processes that should be performed when starting a project (page 43).

   A. Develop Project Charter

   B. Identify Stakeholders

7. List the two processes that should be performed when closing a project (page 43).

   A. Close Project or Phase

   B. Close Procurements

8. The *PMBOK® Guide* consistently uses the verb-noun format to describe each process. If you carefully examine the verb portion of the process name, it should give you an indication that a given process belongs to a certain process group. The noun portion of the process name should also give you a hint of the expected deliverable for that process. For example, the Develop Project Management Plan process indicates that you need to "develop" something after you have started your project—with the "Project Management

Plan" as the expected deliverable. List the 10 common verbs that are used to name the processes within the Planning process group (page 43).

A. Collect

B. Create

C. Define

D. Determine

E. Develop

F. Estimate

G. Identify

H. Perform

I. Plan

J. Sequence

9. List the seven common verbs that are used to name the processes within the Initiating process group (page 43).

A. Acquire

B. Conduct

C. Develop*

D. Direct

E. Distribute

F. Manage

G. Perform*

*** Note that the verbs *develop* and *perform* are also used to describe some of the process names in the Planning process group.**

10. List the six common verbs that are used to name the processes within the Monitoring and Controlling process group (page 43).

A. Administer

B. Control

C. Monitor

D. Perform*

E. Report

F. Verify

*** Note that the verb *perform* is also used to describe some of the process names in the Planning and Initiating process groups.**

Chapter 4: Project Integration Management

11. List at least 10 subsidiary plans of the project management plan (page 82).

- **A.** Scope management plan
- **B.** Requirements management plan
- **C.** Schedule management plan
- **D.** Cost management plan
- **E.** Quality management plan
- **F.** Process improvement plan
- **G.** Human resource plan
- **H.** Communications management plan
- **I.** Risk management plan
- **J.** Procurement management plan

Chapter 5: Project Scope Management

12. List the eight tools and techniques that a project manager can use when collecting requirements (pages 107–109).

- **A.** Interviews
- **B.** Focus groups
- **C.** Facilitated workshops
- **D.** Group creativity techniques
- **E.** Group decision-making techniques
- **F.** Questionnaires and surveys
- **G.** Observations
- **H.** Prototypes

13. List three terms that relate to the work breakdown structure (WBS) (pages 119–120).

- **A.** Work packages
- **B.** Phase
- **C.** Major deliverables

Chapter 6: Project Time Management

14. What are the four types of activity dependencies or logical relationships (page 138)?

A. Finish-to-start (FS)

B. Finish-to-finish (FF)

C. Start-to-start (SS)

D. Start-to-finish (SF)

15. List five estimating tools and techniques (pages 149–151).

A. Expert judgment

B. Analogous estimating

C. Parametric estimating

D. Three-point estimates

E. Reserve analysis

16. What is the PERT formula for three-point estimating (page 150)?

$t_E = (t_O + 4t_M + t_P) / 6$

Chapter 7: Project Cost Management

Define the following Earned Value Management (EVM) acronyms (pages 182–185).

17. What is PV? Planned value

18. What is EV? Earned value

19. What is AC? Actual cost

20. What is SV? Schedule variance

21. What is CV Cost variance

22. What is SPI? Schedule performance index

23. What is CPI? Cost performance index

24. What is EAC? Estimate at completion

25. What is ETC? Estimate to complete

26. What is TCPI? To-complete performance index

27. List the two EVM variance formulas (page 182).
    - A. SV = EV – PV
    - B. CV = EV – AC

28. List the two EVM performance index formulas (page 183).
    - A. SPI = EV / PV
    - B. CPI = EV / AC

29. List the four EVM forecasting formulas for EAC (pages 184–185).
    - A. EAC = AC + bottom-up ETC
    - B. EAC = AC + BAC – EV
    - C. EAC = BAC / cumulative CPI
    - D. EAC = AC + [(BAC – EV) / (cumulative CPI × cumulative SPI)]

30. List the two to-complete performance index formulas (page 185).
    - A. TCPI based on the BAC = (BAC – EV) / (BAC – AC)
    - B. TCPI based on the EAC = (BAC – EV) / (EAC – AC)

Chapter 8: Project Quality Management

Define the following quality-specific acronyms (page 191).

31. What is PDCA? Plan-do-check-act cycle

32. What is TQM? Total quality management

33. What is OPM3? Organizational Project Management Maturity Model

34. What is CMMI? Capability Maturity Model Integration

35. What is COQ? Cost of quality

36. Name the two costs of quality (page 195).
    - A. Cost of conformance
    - B. Cost of nonconformance

37. What are the two categories of the cost of quality for avoiding failures (page 195)?
    - A. Prevention costs
    - B. Appraisal costs

38. What are the two categories of the cost of quality where money must be spent because of failures (page 195)?
    - A. Internal failure costs
    - B. External failure costs

39. List Ishikawa's seven basic tools of quality (pages 208–212).
   A. Cause and effect diagrams
   B. Control charts
   C. Flowcharting
   D. Histogram
   E. Pareto charts
   F. Run charts
   G. Scatter diagrams

Chapter 9: Project Human Resource Management

40. List in order the five stages of developing a team (page 233).
   A. Forming
   B. Storming
   C. Norming
   D. Performing
   E. Adjourning

41. What are the six general techniques for avoiding conflict (page 240)?
   A. Withdrawing/avoiding
   B. Smoothing/accommodating
   C. Compromising
   D. Forcing
   F. Collaborating
   G. Confronting/problem solving

Chapter 10: Project Communications Management

42. What are the five key components of a communication model (page 255)?
   A. Encode
   B. Message and feedback-message
   C. Medium
   D. Noise
   E. Decode

43. List the three broad classifications of communication methods (page 256).

A. Interactive communication

B. Push communication

C. Pull communication

Chapter 11: Project Risk Management

44. List four strategies for negative risks or threats (pages 303–304).

A. Avoid

B. Transfer

C. Mitigate

D. Accept

45. List four strategies for positive risks or opportunities (pages 304–305).

A. Exploit

B. Share

C. Enhance

D. Accept

Chapter 12: Project Procurement Management

46. What are the three major types of contracts (pages 322–324)?

A. Fixed-price contracts (also known as lump sum)

B. Cost-reimbursable contracts (also known as cost plus)

C. Time and materials contracts (T&M)

There are six potential variations of the major types of contracts (pages 322–324).

47. What is FFP? Firm fixed price contracts

48. What is FPIF? Fixed price incentive fee contracts

49. What is FP-EPA? Fixed price economic price adjustment contracts

50. What is CPFF? Cost plus fixed fee contracts

51. What is CPIF? Cost plus incentive fee contracts

52. What is CPAF? Cost plus award fee contracts

Appendix G: Interpersonal Skills

53. List at least five interpersonal skills (page 417).

A. Leadership

B. Team Building

C. Motivation

D. Communication

E. Influencing

Decision making, political and cultural awareness, and negotiation are also interpersonal skills.

54. List at least three team member communication styles (page 419).

A. Directive

B. Collaborative

C. Logical

Explorer can also be added to the list of team member communication styles.

55. What are the two categories of listening techniques (page 419)?

A. Active

B. Effective

56. What are the four basic decision styles (page 420)?

A. Command

B. Consultation

C. Consensus

D. Coin flip (random)

57. What are the four major factors that affect the decision styles (page 420)?

A. Time constraints

B. Trust

C. Quality

D. Acceptance

Define the following acronyms (pages 424–425).

58. What is CCB? Change control board

59. What is CPM? Critical path methodology

60. What is EMV? Expected monetary value

61. What is FMEA? Failure mode and effect analysis

62. What is IFB? Invitation for bid

63. What is LOE? Level of effort

64. What is OBS? Organizational breakdown structure

65. What is PDM? Precedence diagramming method

66. What is QA? Quality assurance

67. What is QC? Quality control

68. What is RACI? Responsible, accountable, consult, and inform

69. What is RAM? Responsibility assignment matrix

70. What is RBS? Risk breakdown structure

71. What is RFI? Request for information

72. What is RFP? Request for Proposal

73. What is RFQ? Request for quotation

74. What is SOW? Statement of work

75. What is SWOT? Strengths, weaknesses, opportunities, and threats

PMI's *Code of Ethics and Professional Conduct*

The page numbers refer to the page number of the corresponding text in the PMI's *Code of Ethics and Professional Conduct*.

76. The PMI's *Code of Ethics and Professional Conduct* applies to which individuals (page 46)?

    A. All PMI members

    B. Nonmembers who hold a PMI certification

    C. Nonmembers who apply to commence a PMI certification process

    D. Nonmembers who serve PMI in a volunteer capacity

77. What two types of responsibility, fairness, and honesty standards were mentioned in the *Code of Ethics and Professional Conduct* (page 46)?

    A. Aspirational conduct

    B. Mandatory conduct

Key Concepts and Processes

Many of these questions discuss topics and concepts presented in Sybex's *PMP: Project Management Professional Exam Study Guide, 6th Edition* (ISBN: 978-1118083215). The page numbers in the answer refer to the page numbers of the corresponding Sybex book.

This section presents questions based on key project management concepts (e.g., projects, programs, portfolios, project management skills, etc.) and various processes from initiating to closing projects.

Projects

78. Projects are temporary initiatives with defined start and end dates (page 3).
79. The concept of incrementally and continually refining the characteristics of a product, service, or result as the project progresses is known as progressive elaboration (page 3).
80. Unlike projects, operations are ongoing and repetitive (page 4).
81. Stakeholders are individuals or groups with a vested interest in the outcome of the project (page 5).
82. The project sponsor is usually an executive within the organization who has authority over the project, from allocating resources to making key decisions (page 5).
83. The project manager is responsible for applying the tools and techniques of project management, and managing the associated resources to ensure that the project objectives will be met (page 7).
84. Programs are groups of related projects (page 8).
85. Portfolios include projects, programs, and other portfolios (page 9).
86. The project management office (PMO) serves as a central oversight in the management of projects and programs within an organization (page 9).
87. What is a key disadvantage for a project manager when working in a functional organization (page 16)?

    They have little to no formal authority.
88. Projectized organizations are considered the opposite of functional organizations (page 18).
89. What are the three common types of matrix structures (page 22)?
    - A. Weak matrix
    - B. Balanced matrix
    - C. Strong matrix

Creating the Project Charter

Fill in the blanks or answer the questions below.

90. Who is not involved in project selection: project manager, project sponsor, customer, or subject matter experts (page 59)?

Project manager

91. <u>Project selection methods</u> are also used to evaluate and choose between alternative ways of performing the project (page 60).

92. What is the discount rate when NPV equals zero (page 66)?

IRR

93. Which project should be chosen: low IRR or high IRR (page 66)?

The project with the higher IRR value

94. The <u>Develop the Project Charter</u> process produces the project charter (page 67).

95. Corporate procedures, policies, standards, and templates are elements of <u>organizational process assets</u> (page 71).

96. Who should be the author of the project charter (page 75)?

Executive manager in the organization

97. Projects that have an NPV <u>greater than</u> zero have a higher likelihood of being approved (page 86).

98. <u>Enterprise environmental factors</u> are external factors that can positively or negatively impact the outcome of the project (page 86).

99. Name at least two needs or demands that necessitate the initiation of a project (page 86).

A. Market demand

B. Organizational need

C. Customer requests

D. Technological advances

E. Legal requirements

F. Ecological impacts

G. Social needs

Developing the Project Scope Statement

Determine if each of the following statements is true or false. If the statement is false, provide the correct answer to the underlined term(s) to make the statement true.

100. The project charter defines how the project will be executed, monitored and controlled, and closed (page 103).

False; project management plan

101. Requirements must be documented, analyzed, and quantified in enough detail that they can be measured once the work of the project begins (page 104).

True

102. Focus Groups comprise cross-functional stakeholders (page 106).

False; facilitated workshops

103. The requirements traceability matrix enables the organization to link each requirement to business and project objective (page 110).

True

104. The project scope management plan documents how the project team will define the project scope, develop the work breakdown structure, and control the changes (page 110).

True

105. The Create WBS process enables the project team to subdivide the project into smaller and more manageable components (page 123).

True

106. Rolling wave planning is a process of elaborating deliverables, project phases, or subprojects in the WBS (page 130).

True

107. The scope baseline includes the scope statement, the WBS, and the project schedule (page 134).

False; project scope statement, the WBS, and the WBS dictionary

108. The scope management plan is a subsidiary plan of the project charter (page 140).

False; project management plan

109. The lowest level of any WBS is called a work package (page 140).

True

Creating the Project Schedule

Match the terms below with their correct descriptions. Write the letter of the correct description in the space provided (pages 153, 156–159, 176, 179 and 192).

110. Analogous estimating I

111. Define Activities process B

112. Discretionary dependency A

113. Finish-to-start E

114. Lag C

115. Lead H

116. Monte Carlo G

117. One standard deviation D

118. PDM J

119. Start-to-finish F

Descriptions

A. Creates arbitrary total float values

B. Decomposes the work packages into schedule activities

C. Delays successor activities

D. Gives a probability of about 68 percent

E. Most commonly used dependency in the PDM method

F. Rarely used dependency in the PDM method

G. Shows probability of all the possible project completion dates

H. Speeds up successor activities

I. Uses expert judgment and historical information

J. Uses only one time estimate to determine duration

Developing the Project Budget

Fill in the blanks or answer the questions below.

120. The cost management plan is established using the WBS and its associated control accounts (page 203).

121. The control account also has a unique identifier that's linked to the organization's accounting system, sometimes known as a chart of accounts (page 203).

122. Scope definition should be completed early in the project to help estimate the costs (page 205).

123. The following three additional tools can be used to predict the potential financial performance of the project's product, service, or result: return on investment, discounted cash flow, and payback analysis (page 207).

124. A project's cost performance baseline normally has a/an S curve shape (page 213).

125. Effective communication entails providing the information in the correct format to the right audience in a timely manner (page 216).

126. What is the communication channel formula (page 217)?

C = [n (n −1)] / 2 where n is the number of stakeholders (participants) in the communication model.

127. The lines that connect the communication nodes are also known as lines of communication (page 217).

128. The communications management plan addresses the communication needs of the project stakeholders (page 221).

129. What is a PMB (page 225)?

Performance measurement baseline

Risk Planning

Determine if each statement below is true or false. If the statement is false, provide the correct answer to the underlined term(s) to make the statement true.

130. Not all risks are bad (page 236).

True

131. The Plan Risk Management process serves as the foundation for subsequent risk processes (page 237).

True

132. RBS is an acronym for requirements budgeting system (page 242).

False; risk breakdown structure

133. Catastrophic risks are also known as residual risks (page 243).

False; force majeure

134. Information gathering is a type of brainstorming (page 247).

False; nominal group technique

135. The fishbone diagram is also known as the Ishikawa diagram (page 249).

True

136. Probability is the likelihood that an event will occur (page 255).

True

137. Sensitivity analysis data can be displayed using a sensitivity diagram (page 264).

False; tornado diagram

138. The sum of the probabilities for each node in a decision tree is 1 (page 266).

True

139. The following strategies can be used to deal with negative risks or threats: avoid, transfer, mitigate, and share (page 269).

False; avoid, transfer, mitigate, and accept

Planning Project Resources

Match the terms below with their correct descriptions. Write the letter of the correct description in the space provided (pages 292–293, 295, 297, 300, 304–305, 309, and 311).

140. Collective bargaining agreements C

141. Firm fixed-price G

142. FP-EPA economic adjustment F

143. Make-or-buy analysis D

144. Organization breakdown structure I

145. Project calendars J

146. Regulation H

147. Source selection criteria E

148. Teaming agreement B

149. Time and materials contracts A

Descriptions

A. Crosses between fixed-price and cost-reimbursable contracts

B. Defines the contractual agreements between multiple parties

C. Defines the organization's contractual obligations to employees

D. Facilitates the decision-making process whether to build or purchase products and services that are needed in the project

E. Helps organizations choose a vendor from among the proposals received

F. Mandates linking it to a known financial index

G. Places the bulk of the risks on the seller

H. Requires compliance and are almost always imposed by governments or institutions

I. Shows the departments, work units, etc. within an organization

J. Shows the working and nonworking days in the project

Developing the Project Team

Fill in the blanks or answer the questions below.

150. Which comes first: validated defect repair or defect repair (page 345)?

Defect repair

151. Which comes first: norming or storming (page 353)?

Storming

152. Which comes first: storming or forming (page 353)?

Forming

153. To achieve co-location, the project manager may set aside a common meeting area, which is sometimes called the war room (page 356).

154. Motivation can be extrinsic or intrinsic (page 357).

155. Which one is higher in the hierarchy of needs: social needs or basic physical needs (page 359)?

Social needs

156. Which one is higher in the hierarchy of needs: safety and security needs, or self-actualization (page 359)?

Self-actualization

157. Expectation Theory: positive outcomes drive motivation (page 361).

158. Which type of power should be used as a last resort: legitimate or punishment (page 364)?

Punishment

159. What are the most important skills that a project manager should possess (page 376)?

Communication skills

Coordinating Procurements and Sharing Information

Fill in the blanks or answer the questions below.

160. Which proposal evaluation technique assigns numerical weights to evaluation criteria: weighting systems or screening systems (page 390)?

Weighting systems

161. Which proposal evaluation technique uses predefined performance criteria: weighting systems or screening systems (page 390)?

Screening systems

162. Which proposal evaluation technique relies on information about the sellers: screening systems or seller rating systems (page 390)?

Seller rating systems

163. Independent estimates, sometimes conducted by the procurement department to compare the proposal costs against the vendor costs, are also known as should-cost estimates (page 392).

164. What tactic is used during contract negotiation when one party tries to discuss an issue that is no longer relevant (page 393)?

Fait accompli

165. Which contract life-cycle stage comes first: solicitation or requisition (page 396)?

Requisition

166. Which contract life-cycle stage comes first: requirement or award (page 396)?

Requirement

167. Quality audits are independent reviews performed by trained auditors or third-party reviewers (page 398).

168. The action item log helps promote communication with stakeholders (page 407).

169. Contested changes can take form of disputes, claims, or appeals (page 412).

Measuring and Controlling Project Performance

Fill in the blanks or answer the questions below.

170. A work authorization system is a subsystem of the project management system (page 425).

171. To get paid, vendors issue seller invoices (page 428).

172. When a seller submits an invoice, the payment system is used to issue payments (page 430).

173. What is ADR (page 430)?

Alternative dispute resolution

174. Time-series methods use historical data to predict future performance (page 434).

175. Judgmental method is a forecasting method that uses opinions, intuition, and estimates to predict future outcomes (page 435).

176. Who is usually the expediter of project status meeting (page 435)?

Project manager

177. What do you need to do to introduce modifications to the project (page 438)?

Submit a change request

178. Is the change control system a subset of the configuration management system (page 441)?

Yes

179. Which one is the preferred approach for capturing change requests: written or verbal (page 448)?

Written

Controlling Work Results

Determine if each statement below is true or false. If the statement is false, provide the correct answer to the underlined term(s) to make the statement true.

180. Planned value is the same as budgeted cost of work scheduled (page 465).

True

181. Actual cost is the same as actual cost of work performed (page 466).

True

182. Earned value is the same as actual cost of work performed (page 466).

False; budgeted cost of work performed

183. Given an EV of 350 and AC of 300, the CV is 650 (page 467).

False; CV = EV – AC = 350 – 300 = 50

184. A CPI greater than 1 is good (page 468).

True

185. EAC = AC + BAC – EV (page 469).

True

186. VAC = BAC – EAC (page 472).

True

187. CV = EV – PV (page 473).

False; SV

188. A control chart has lower and upper control limits (page 481).

True

189. Ishikawa is credited with the 80/20 Rule (page 482).

False; Pareto

190. A scatter diagram's two variables are dependent and independent variables (page 484).

True

Closing the Project and Applying Professional Responsibility

Match the terms below with their correct descriptions. Write the letter of the correct description in the space provided (pages 508, 513, 517, 522, and 541).

191. Addition, starvation, integration, and extinction J

192. Aspirational and mandatory I

193. Closed procurements F

194. Lessons Learned D

195. Organization process updates H

196. PMI's *Code of Ethics and Professional Conduct* E

197. Procurement audit A

198. Procurement early termination C

199. Project management plan G

200. Responsibility, respect, fairness, and honesty B

Descriptions

A. Applicable to both buyer and vendor

B. Areas of professional responsibility

C. By agreement, via default or for cause

D. Captures what went well in the project and what should be avoided for the next project

E. Ethical standards for PMPs

F. Formal acceptance and closure of the procurement

G. Input of Close Procurements process

H. Output of the Close Project or Phase process

I. Standards of professional responsibility

J. Types of project endings

Practice Test A

1. Which knowledge area unifies the various processes and activities within the project management process groups?
   - A. Project Integration Management
   - B. Project Scope Management
   - C. Project Quality Management
   - D. Project Human Resource Management

2. Over the years of working together, a senior program manager and a project manager developed a strong professional relationship. The project manager will be starting a large project within the next few weeks, so he asked the senior program manager for advice on how to deal with a new vendor (one that the company has never used). Without hesitation, the senior program manager told the project manager just to use the same contract that the company has used for smaller vendors. What should the project manager do next?
   - A. Get a copy of the contract for small vendors and then ask the new vendor to sign it.
   - B. Ask the new vendor for a copy of its standard contract and then sign the contract.
   - C. Seek advice from other project managers and then personally prepare a new contract.
   - D. Solicit inputs from the legal and procurement departments and then proceed accordingly.

3. Which project management process should a project manager perform to get a formal authorization for a project or a project phase?
   - A. Develop Project Management plan
   - B. Develop Project Charter
   - C. Perform Integrated Change Control
   - D. Direct and Manage Project Execution

4. Project managers need to perform several project management processes as part of the Project Integration Management knowledge area. During project initiation, the project manager will need to ______________.
   - A. Develop the Project Management plan
   - B. Manage stakeholder expectations
   - C. Develop the project charter
   - D. Perform integrated change control

5. You have been asked to manage the change control process of a very large government project. Some of your responsibilities include reviewing, approving, and managing the changes. The operations manager prepared a list of expected deliverables from your project. She asked you to cross out deliverables that are not applicable to your role. From the following list, which deliverable should you remove?

   A. Change request status updates

   B. Project management plan updates

   C. Project document updates

   D. Organizational process assets updates

6. Which of the following tools or techniques is used in all of the Project Integration Management knowledge area processes?

   A. Expert judgment

   B. Project management information system

   C. Change control meetings

   D. Facilitated workshops

7. The Project management plan enables the project manager to ________________.

   A. Monitor and control project work

   B. Develop a project scope statement

   C. Direct and manage project execution

   D. Implement integrated change control

8. A project manager gathered the following inputs: project management plan, accepted deliverables, and organizational process assets. If the project manager will use expert judgment to transform these inputs into outputs, she is likely performing which Project Integration Management knowledge area process?

   A. Direct and Manage Project Execution

   B. Develop Project Charter

   C. Perform Integrated Change Control

   D. Close Project or Phase

9. Who authorizes the project charter?

   A. Project initiator or sponsor

   B. Program manager and project manager

   C. Functional or department manager

   D. Finance managers and chartered accountants

10. Which item from the following list may be initiated prior to the development of the project charter?
    - A. Preliminary study and project management plan
    - B. Needs assessment and business case
    - C. Market analysis and detailed cost estimates
    - D. Project scope statement and work breakdown structure

11. You are working as a project management consultant for a regional financial institution. One of your key deliverables is a project charter template for the project management office. What information should you include, via direct or indirect references, in the project charter template?
    - A. Project justification, success criteria, and summary milestone schedule
    - B. Product objectives and product scope description
    - C. Project requirements and deliverables, project constraints, and project assumptions
    - D. Project deliverables and project exclusions

12. Your company has asked you to prepare a project charter for an external manufacturing customer. The customer expects you to deliver the installation, configuration, and testing of the new state-of-the-art assembly line that was developed by your company. What key input will you need in order to prepare the project charter?
    - A. Project scope statement
    - B. Work performance information
    - C. Project management plan
    - D. Contract

13. The project statement of work (SOW) serves as one of the inputs when developing a project charter. What type of information should a project manager look for in a SOW?
    - A. Business need, product scope description, and strategic plan
    - B. Organizational structure, industry standards, and authorization system
    - C. Stakeholder risk tolerances, commercial databases, and marketplace conditions
    - D. Infrastructure, human resources, and organizational culture

14. In order for a project to be successful, the project manager must consider various enterprise environmental factors. Which of the following lists contain only enterprise environmental factors?
    - A. Business need, product scope description, Lessons Learned knowledge base, and strategic plan
    - B. Government standards, infrastructure, and project management information systems
    - C. Standardized cost estimating data; strengths, weaknesses/limitations, opportunities, and threats (SWOT) analysis; and industry risk study information
    - D. Automated tool suite, electronic document repository, and configuration management system

15. Submitting time reports, reviewing disbursements, assigning codes of accounts, and using contract provisions are examples of ______________.

A. Project closure guidelines

B. Risk control procedures

C. Financial control procedures

D. Organizational standard processes

16. What are the two categories of organizational process assets?

A. Organization communication requirements and work authorization procedures

B. Organizational standard processes and financial control procedures

C. Issue and defect management procedures, and risk control procedures

D. Organizational processes and procedures, and corporate knowledge base

17. You have been assigned to manage a medical information system project for a major client of your company. In the past, your company has successfully delivered several similar projects for the same client. Which of the following organizational corporate knowledge bases would you review prior to the preparation of the project charter?

A. Organizational standard processes, standardized guidelines, templates, project customization guidelines, project closure guidelines, and issue and defect management procedures

B. Process measurement database, historical information, Lessons Learned, issue and defect management database, financial database, and configuration management knowledge base

C. Financial control procedures, issue and defect management procedures, change control procedures, risk control procedures, and organizational processes and procedures

D. Proposal evaluation criteria, security requirements, safety and health policy, project management policy, defect identification procedures, and action item tracking

18. The project statement of work (SOW) ______________.

A. Is a narrative description of products or services to be delivered by the project

B. Is a document that formally initiates the project and identifies the project manager

C. Is developed from the project scope baseline and defines that portion of scope to be included in the contract

D. Describes, in detail, the project's deliverables and the work required to create those deliverables

19. Which project management process is used to define and document the high-level requirements and description of the project?

A. Develop Project Charter

B. Develop Project Management plan

C. Develop Project Scope Statement

D. Monitor and Control Project Work

20. Luigi took over a project from another project manager who abruptly left the company. No one knows the state of the project; however, Luigi found some documents that contain the project high-level description, summary milestone schedule, and name and authority of the person authorizing the project. Most likely, the document that Luigi found is the ______________.

A. Project charter

B. Project management plan

C. Project statement of work

D. Project scope statement

21. To complete the project charter, a project manager will need to use information provided by the ______________.

A. Project initiator or sponsor

B. Project scope department

C. Enterprise environmental factors

D. Organizational process assets

22. What inputs are needed to develop the project charter?

A. Contract, project statement of work, enterprise environmental factors, and organizational process assets

B. Project scope statement, requirements documentation, and approved changed requests

C. Project management plan, work performance information, and rejected change requests

D. Project management processes, enterprise environmental factors, and organizational process assets

23. The project management team has selected the project management processes that will be used in an oil mining project. In which document will a project sponsor be able to find such information?

A. Project charter

B. Project scope statement

C. Project management plan

D. Project statement of work

24. Selecting buyers, managing risks, managing sellers, adapting approved changes, collecting project data, implementing standards, training the project team, and performing the activities to achieve the project objectives are actions associated with which of the following project management processes?

A. Verify and Control Scope

B. Control Cost and Schedule

C. Direct and Manage Project Execution

D. Administer Procurements and Monitor Risks

25. You are managing a construction project that builds roads and bridges. In addition to tangible deliverables, what intangible deliverables can the project produce or deliver as part of project execution?

A. Staff skills

B. Highway lighting

C. Traffic signs

D. None—all project deliverables must be tangible

26. Collecting the project data and facilitating the forecasting of cost and schedule are actions performed by the project manager and the project management team during which of the following project process groups?

A. Initiating

B. Planning

C. Executing

D. Monitoring and Controlling

27. Approved _______________ are documented, authorized directions required to bring expected future project performance into compliance with the project management plan.

A. Corrective actions

B. Preventive actions

C. Change requests

D. Defect repairs

28. Which input to the Direct and Manage Project Execution process contains the documented and authorized changes to adjust the scope of the project?

A. Approved corrective actions

B. Approved preventive actions

C. Approved change requests

D. Approved defect repair

29. ______________ define(s) a process that guides a project team in executing the project management plan.

A. Project management information system

B. Work performance information

C. Rejected change requests

D. Expert judgment

30. You are working on a secret project for the military. Because of a recent military order, the project will need to include extensive changes to the information system, internal procedures, and standardized forms. These changes were considered and accommodated during project initiation and planning. From here on, what other outputs are you expected to deliver?

A. Deliverables, corrective actions, and defect repair

B. Recommended corrective actions, forecasts, and requested changes

C. Approved change requests, rejected change requests, and deliverables

D. Contract closure procedure, along with final product, service, or result

31. Which description *best* characterizes work performance information?

A. Enumerates changes to the project scope, lists potential changes to policies, addresses mandatory changes, and adjusts the overall project budget as needed

B. Indicates deliverable status, schedule progress, and costs incurred

C. Completes the project upon delivery, performs a service as outlined in the project management plan, and helps generate results

D. Corrects future project performance, helps reduce the negative consequences of project risks, and proves that change requests were implemented

32. The Monitor and Control Project Work process examines processes during which of the following process groups?

A. Initiating, Planning, Executing, and Closing

B. Initiating, Planning, and Monitoring and Controlling

C. Monitoring and Controlling, and Closing

D. Executing, and Monitoring and Controlling

33. You have been assigned to a program in which you'll work with academic researchers. The program itself involves selecting promising research projects, matching available funding to each research proposal, monitoring the progress of each research project, and canceling research projects that cannot meet predefined milestones and benchmarks. During project execution, the earned value management technique will serve as the primary cost control assessment tool to determine whether a project is progressing according to the project management plan.

What must you do as a project manager to facilitate the application of the earned value technique on various research projects?

A. Compare actual project performance to the project management plan

B. Assess project performance to determine if corrective actions are needed

C. Monitor the implementation of approved changes and manage project risks

D. Establish project baselines, and capture overall progress and variances

34. You have been assigned to a program in which you'll work with academic researchers. The program itself involves selecting promising research projects, matching available funding to each research proposal, monitoring the progress of each research project, and canceling research projects that cannot meet predefined milestones and benchmarks. During project execution, the earned value technique will serve as the primary cost control assessment tool to determine whether a project is progressing according to the project management plan.

Research Project A reported an earned value of $500 and an actual cost of $505. Research Project B reported an earned value of $455 and an actual cost of $450. Research Project C submitted a cost performance index of 1.08. Research Project D achieved a cost performance index of 0.94. Which research project is the *most* cost efficient?

A. Research Project A

B. Research Project B

C. Research Project C

D. Research Project D

35. You have been assigned to a program in which you'll work with academic researchers. The program itself involves selecting promising research projects, matching available funding to each research proposal, monitoring the progress of each research project, and canceling research projects that cannot meet predefined milestones and benchmarks. During project execution, the earned value technique will serve as the primary cost control assessment tool to determine whether a project is progressing according to the project management plan.

You need to determine if corrective actions are required for Research Project E. Given an earned value of $747 and an actual cost of $691, calculate the cost variance.

A. $56

B. –$56

C. 1.08

D. 0.93

36. Project managers need to perform change control from the start through the end of the project. Which configuration management activity is *not* part of the Perform Integrated Change Control process?

A. Configuration identification

B. Architecture configuration and control

C. Configuration status accounting

D. Configuration verification and auditing

37. Documented change requests *must* be either ____________ or ____________ by the change control board.

A. Accepted, deferred

B. Reviewed, deferred

C. Reviewed, accepted

D. Accepted, rejected

38. The change control board on Paul's marketing campaign project approved the inclusion of television advertisements in addition to radio, print, and online advertisements. As part of integration change control, which documents should Paul review to determine the impact of this change and whether updates are required?

A. Project management plan and organizational process assets

B. Organizational process assets and project scope statement

C. Project management plan

D. Administrative closure procedure and contract closure procedure

39. Which of the following is *not* a Perform Integrated Change Control process tool or technique?

A. Earned value technique

B. Change control meetings

C. Change control board

D. Expert judgment

40. If you are working on a multiphase project, when is it appropriate to perform the tasks associated with the Close Project or Phase process?

A. At a specific milestone

B. At the end of each phase

C. At the end of the calendar year

D. At the end of the fiscal year

41. You have reached the end of your project, so you have started to collect project documents and gather Lessons Learned. In addition, you have initiated the archiving of important project information. Which process are these tasks performed in?

A. Record retention

B. Performance review

C. Close procurements

D. Organizational process assets updates

42. You were working on a project that consisted of five major phases: analysis, design, development, testing, and implementation. The testing phase failed to meet the project sponsor's expectations. Because of that failure, the sponsor canceled the project and you were asked to perform a closing procedure. What should you do next?

A. Ask the sponsor to reconsider and try adjusting the project scope statement

B. Make alterations and retest in the hope of meeting the sponsor's expectations

C. Settle and close any contractual claims associated with the project

D. Collect project files and documents and begin gathering Lessons Learned

43. Which of the following are inputs to the Close Project or Phase process?

A. Project management plan, accepted deliverables, and organizational process assets

B. Contract (when applicable), enterprise environmental factors, and organizational process assets

C. Project charter, enterprise environmental factors, and organizational process assets

D. Project scope statement, enterprise environmental factors, and organizational process assets

44. Expert judgment is a tool and technique used by which processes?

A. Develop Project Charter, and Direct and Manage Project Execution

B. Develop Project Charter, Develop Project Management Plan, Direct and Manage Project Execution, Monitor and Control Work, Perform Integrated Change Control, and Close Project or Phase

C. Direct and Manage Project Execution, and Monitor and Control Project Work

D. Perform Integrated Change Control, and Direct and Manage Project Execution

45. Project audits, transition criteria, and previous project performance information can be found as part of (the) _______________.

A. Project management plan

B. Enterprise environmental factors

C. Organizational process assets

D. Contract documentation

46. Six months after you formally closed an office relocation project, you receive a phone call from a collection agency indicating that your company refuses to pay the invoices of one of the vendors. What would be the *best* course of action for you to take?

A. Revisit the Close Procurements process of the project in question

B. Refer the collection agency to the accounts payable department

C. Forward the phone call to your company's legal department

D. Terminate the telephone call because the project is no longer your responsibility

47. You spent one week updating the project files, project closure documents, and historical information. These project documents collectively are part of (the) ______________.

A. Configuration management system

B. Organizational process assets

C. Work performance information

D. Enterprise environmental factors

48. Project closure documents are produced as part of the Close Project or Phase process. Which of the following statements is true about project closure documents?

A. Lessons Learned are transferred to the knowledge database for future use by other projects.

B. Project closure documents formally indicate the official acceptance of project deliverables.

C. Project closure documents formally transfer the project deliverables to the operations group.

D. Project closure documents mark the end of a project phase as well as the end of the project.

49. You have been assigned to a project to initiate, plan, and execute a national convention for travel agents. One of the immediate tasks that you need to perform is to create and define the work breakdown structure (WBS). The creation of the WBS is part of which Knowledge Area process?

A. Project Integration Management

B. Project Scope Management

C. Project Time Management

D. Project Cost Management

50. How are the Define Scope and Create WBS processes different?

A. The outputs of the Create WBS process serve as inputs to the Collect Requirements process; outputs of the Create WBS can be used as inputs to Define Scope process.

B. Major project deliverables are subdivided during the Define Scope process; the major project deliverables are then created and defined as part of the Create WBS process.

C. A detailed description of the project is developed during the Define Scope process, and the Create WBS process subdivides major deliverables into smaller components.

D. There are no differences between these two processes—both are parts of Project Scope Management, and they interact with each other to produce the WBS.

51. Which of the following are some of the tools and techniques of the Define Scope process?

A. Product analysis, alternatives identification, and facilitated workshops

B. Decomposition

C. Interviews, focus groups, questionnaires, and surveys

D. Variance analysis

52. Brainstorming, the Delphi technique, and mind mapping are specific methods for which Collect Requirements tool and technique?

A. Observation

B. Facilitated workshops

C. Group decision-making techniques

D. Group creativity techniques

53. You have been assigned to a project to develop a new bicycle that will use the latest technologies available in the market today. You have started working on the work breakdown structure (WBS) by listing "bicycle" on the first level. Items on the second level include "frameset," "crank set," "wheels," "braking system," and "shifting system." What items would be appropriate at the third level, under "wheels"?

A. Front wheel and rear wheel

B. Concept and design

C. Assembly and testing

D. Specification and testing

54. ______________ is the only tool and technique in the Create WBS process.

A. Inspection

B. Decomposition

C. Variance analysis

D. Facilitated workshops

55. The ________________ hierarchically depicts the project structure in order to properly relate the work packages to the performing organizational units.

A. Work breakdown structure

B. Organizational breakdown structure

C. Risk breakdown structure

D. Resource breakdown structure

56. Which process requires obtaining formal acceptance of the completed project scope and associated deliverables from the stakeholders?

A. Collect Requirements

B. Define Scope

C. Verify Scope

D. Control Scope

57. As part of the Verify Scope process, the project team must measure, examine, and verify the project deliverables to ensure adherence to the requirements. Which tool or technique would be *most* appropriate in performing these tasks?

A. Product analysis

B. Decomposition

C. Inspection

D. Variance analysis

58. Which of the following statements is true about the Control Scope process?

A. The following are examples of inputs to the Control Scope process: project management plan, work performance information, requirements documentation, and organizational process assets.

B. The Control Scope process uses the following tools and techniques: product analysis, alternative identification, expert judgment, and stakeholder analysis.

C. The following are examples of outputs from the Control Scope process: project scope management plan, performance reports, and work performance information.

D. The Control Scope process produces the following outputs: accepted deliverables, requested changes, and recommended corrective actions.

59. Project performance measurements are used to assess the ____________.

A. Magnitude of variation

B. Effectiveness of the team

C. Experience of the project manager

D. Conformance to requirements

60. As a result of variance analysis, the scope baseline or other parts of the project management plan may need to be modified. The results of this are called ____________.

A. Change requests

B. The change control system

C. The configuration management system

D. A requirements traceability matrix

61. Which Project Scope Management process may introduce updates to the work performance measurements, organizational process assets, and project management plan?

A. Collect Requirements

B. Define Scope

C. Verify Scope

D. Control Scope

62. Within the Project Scope Management knowledge area, the scope baseline will be created and may be updated by various processes. From the following list, which processes create and update scope baseline?

A. Collect Requirements and Define Scope

B. Define Scope and Create WBS

C. Create WBS and Control Scope

D. Verify Scope and Control Scope

63. You are trying to determine how to deliver a marketing campaign project on time. You considered using the precedence diagramming method (PDM) to place the project activities in logical order. Which process would normally use the PDM technique?

A. Define Activity

B. Sequence Activity

C. Estimate Activity Resource

D. Develop Schedule

64. In one of your projects, you were assigned to decompose the branches of the WBS down to the work package level. You were able to successfully break down four out of the five branches of the WBS. Unfortunately, the project scope did not provide enough details for you to decompose the fifth branch of the WBS. What should you do next?

A. Ask the sponsor to change the project scope so that you can continue

B. Document the insufficient definition of the WBS branch as a project risk

C. Make reasoned assumptions about the WBS branch so as not to delay the project

D. Use the branch to plan at a higher level of the WBS

65. You are sequencing the activities of an instructor-led course development project. Which of the following statements is true about the tools and techniques used in activity sequencing?

A. The precedence diagramming method (PDM) uses nodes to represent activities, and arrows connect the nodes to show the dependencies.

B. Leads and lags allows for delays and acceleration of successor activities, respectively.

C. The precedence diagramming method (PDM) is used for dependency determination.

D. The two main types of dependency determination are mandatory dependencies (hard logic) and discretionary dependencies (soft logic).

66. The precedence diagramming method (PDM) uses different types of dependencies or precedence relationships. What is the most commonly used type of precedence relationship?

A. Finish-to-start

B. Finish-to-finish

C. Start-to-start

D. Start-to-finish

67. Project managers use various tools and techniques to estimate the required resources (people, equipment, and material) in a project. Which statement is correct about the Estimate Activity Resources process?

A. The Estimate Activity Resources process produces the following outputs: activity list, activity attributes, and resource availability.

B. The Estimate Activity Resources process is closely coordinated with the Estimate Costs process.

C. The Estimate Activity Resources process requires the activity resource requirements and activity attributes as inputs.

D. The Estimate Activity Resources process is closely coordinated with the Develop Schedule process.

68. Calculate the overall project duration for Project X by using the following table.

WBS	Task Name	Duration	WBS Predecessor
1	Project X		
1.1	Start	0 Day	
1.2	A	2 Days	1.1
1.3	B	3 Days	1.1
1.4	C	6 Days	1.1
1.5	D	7 Days	1.2
1.6	E	8 Days	1.5
1.7	F	3 Days	1.3
1.8	G	4 Days	1.7, 1.4
1.9	H	1 Day	1.6, 1.7
1.10	I	2 Days	1.8
1.11	J	6 Days	1.9, 1.10
1.12	End	0 Day	1.11

A. 8 days

B. 12 days

C. 24 days

D. 42 days

69. Calculate the expected (E) value given the following three point estimates: optimistic (O) = 8, most likely (M) = 14, and pessimistic (P) = 16.

A. 12.67

B. 13.33

C. 14.00

D. 16.00

70. Identify the critical path of Project X based on the following table.

WBS	Task Name	Duration	WBS Predecessor
1	Project X		
1.1	Start	0 Day	
1.2	A	2 Days	1.1
1.3	B	3 Days	1.1
1.4	C	6 Days	1.1
1.5	D	7 Days	1.2
1.6	E	8 Days	1.5
1.7	F	3 Days	1.3
1.8	G	4 Days	1.7, 1.4
1.9	H	1 Day	1.6, 1.7
1.10	I	2 Days	1.8
1.11	J	6 Days	1.9, 1.10
1.12	End	0 Day	1.11

A. A, B, C, and J

B. A, B, C, and I

C. C, D, E, and J

D. A, D, E, H, and J

71. You need to present the monthly status of your construction project to the steering committee. The chair of the committee indicated that she is interested only in the start and finish of key project deliverables. What format should you use to present the project's monthly status?

A. Project schedule network diagrams

B. Bar charts

C. Milestone charts

D. Project schedule

72. Which of the following are outputs of the Develop Schedule process?

A. Schedule data, schedule baseline (updates), and performance measurements

B. Project schedule, schedule baseline, schedule data, and project document updates

C. Requested changes, recommended corrective actions, and organizational process assets (updates)

D. Activity list (updates), activity attributes (updates), and project management plan (updates)

73. Which of the following statements is correct about the Control Schedule process?

A. The Control Schedule process enables the project manager to evaluate, influence, and manage changes in the project schedule.

B. The Control Schedule process occurs prior to the development of the activity list, activity attributes, and milestone list.

C. The Estimate Activity Resources and Develop Schedule processes occur after the Control Schedule process.

D. The Control Schedule and Control Cost processes are the most important processes in the Executing process group.

74. The Control Schedule process is performed as part of the ______________ process group.

A. Executing

B. Monitoring and Controlling

C. Planning

D. Initiating

75. You are working on a network installation project for a major telecommunications company. What do you need in order to determine if updates are required in the organizational process assets, project documents, and project management plan?

A. Organizational process assets, project scope statement, and activity list

B. Activity attributes, project schedule network diagrams, and activity resource requirements

C. Project management plan, schedule baseline, work performance information, and organizational process assets

D. Resource calendars, activity duration estimates, and enterprise environmental factors

76. The project sponsor asked you to audit the schedule of a project in a chocolate factory. The project manager provided you with an earned value of $1,234. The number appeared to be correct. The project baseline shows a planned value of $1,540. What should you do next?

A. Do not trust the project manager; calculate the earned value yourself to be sure.

B. Calculate the SPI and tell the project manager that no corrective actions are required.

C. Present your findings and ask the project sponsor to consider corrective actions.

D. Confirm the earned value and thank the project manager for doing a great job.

77. What should a project manager do if a project has an SPI of 1.29?

A. Confirm that the EV and PV are correct just to be sure.

B. Introduce corrective actions to bring the project back on track.

C. Ask the project team to crash the overall project schedule.

D. Perform resource smoothing to avoid working overtime.

78. What is the schedule variance of a project if the earned value is 888 and the planned value 999?

A. –111.000

B. 111.000

C. 0.889

D. 1.125

79. The three Project Cost Management knowledge area processes are ______________.

A. Estimate Costs, Determine Budget, and Control Costs

B. Determine Budget, Performance Reporting, and Close Project

C. Define Scope, Estimate Costs, and Performance Reporting

D. Plan Procurements, Control Costs, and Administration Procurements

80. The accuracy of estimates will get refined as the project progresses because more detailed information will become available. You were asked to provide cost estimates during the initiation phase of a parasailing expedition project. This is the first time your company has undertaken such a project. At this point in the project, your cost estimates will have a rough order of magnitude (ROM) in what range?

A. –5 to +5%

B. –10 to +10%

C. –25 to +25%

D. –50 to +50%

81. You asked several construction companies to provide estimates for how much it will cost to finish your basement. They asked you for the square footage of the house and whether you want the floor to be finished in carpet, tiles, or hardwood. The construction estimator paused briefly, entered some numbers using a simple calculator, and then gave you a cost estimate. The construction estimator *most likely* used what type of estimating technique?

A. Analogous estimating

B. Bottom-up estimating

C. Parametric estimating

D. Vendor bid analysis

82. Project managers may use analogous or parametric estimating to predict the overall cost of a project. Estimates derived using analogous or parametric models are considered accurate and reliable when ______________.

A. The model is prepared by an expert, a certified project management professional collects the parameters, and the calculations allow for subjective measures (for example, aesthetics).

B. The project being estimated is almost exactly the same as previous projects not just in appearance but also in actual fact (for example, a similar four-bedroom house in the same area).

C. The assumptions, constraints, and limitations of the three point estimates (optimistic, most likely, and pessimistic) have been reviewed, discussed, and finalized by the project team.

D. The parameters are quantifiable, the calculations are based on precise historical information, and the model can be used on projects regardless of size (for example, small vs. large project).

83. After performing the Control Costs process, which of the following may be updated?

A. Budget forecasts, project management plan, and project documents

B. Project funding requirements and performance reports

C. Work performance information and organizational process assets

D. Performance reports and organizational process assets

84. Given BAC = 500, EV^C = 400 and CPI^C = 1.3, forecast the ETC based on atypical variances.

A. 1.25

B. 0.8

C. 650

D. 100

85. Which Project Quality Management process uses the cost of quality (COQ) technique?

A. Plan Quality

B. Quality Function Deployment

C. Quality Control

D. Quality Audit

86. Quality management approaches can be divided into two categories: proprietary and nonproprietary. The following quality management approaches are considered proprietary *except* for _______________.

A. Deming

B. Juran

C. Crosby

D. COQ

87. _______________ are also known as the cost of poor quality.

A. Failure costs

B. Internal failures

C. External failures

D. Cost of quality

88. From a project perspective, how are quality assurance and quality control different?

A. Quality assurance deals with auditing the quality requirements and results, whereas quality control monitors and records results of executing the quality activities.

B. At the project level, quality assurance and quality control are essentially the same—they differ only if used in a nonproject environment.

C. Quality assurance and quality control share similar inputs, but they use different tools and techniques to generate project outputs.

D. Depending on the process defined in the quality management plan, quality assurance is mandatory and quality control is optional.

89. The process improvement plan details the step-by-step activities required to identify, analyze, and remove non–value-added activities in order to increase customer value. Such a plan will analyze the following, *except* for _______________.

A. Process boundaries

B. Process configuration

C. Target for improved performance

D. Configuration management

90. Quality planning tools and techniques, quality audits, and process analyses are used in the _______________ process to convert inputs into outputs.

A. Plan Quality

B. Perform Quality Assurance

C. Perform Quality Control

D. Project Quality Management

91. As a project manager of a major environmental project, you met with local government agencies to address their concerns about the impact of the project on the migration patterns of birds. After the meeting, three change requests were documented and subsequently approved. After reviewing the change requests in detail, however, your team realized that the three change requests cannot be implemented without an additional environmental study. What should you do next?

A. Include the environmental study in the project scope because it is necessary to complete the documented and approved change requests.

B. Check the project's contingency reserve to determine if the cost of the environmental study can be included without additional funding.

C. Ask the team to immediately work on the environmental study so as not to delay the implementation of the three change requests.

D. Do not proceed or implement the environmental study because it is not documented and it has not been approved for implementation.

92. As part of your company's ISO-compliant project management processes, a project quality auditor visits all large projects at least once a month. The project auditor visited your project for five consecutive months and found it to be in compliance with the company's project management processes, but the project auditor did not show up last month. What would be an appropriate course of action?

A. Wait until the next month because the audit is not adding any value to the project.

B. Do not worry about it because the project has been compliant in the past.

C. Contact the project quality auditor and ask him/her to perform the monthly project audit as planned.

D. Report the project quality auditor to the quality management office immediately.

93. ______________ is an assessment performed by an unbiased individual or entity to determine if project activities conform to the organization and project policies and procedures.

A. Quality audit

B. Configuration management system

C. Change control system

D. Process analysis

94. Project managers plan quality activities to ensure that the project will meet the requirements for which it was undertaken. Such activities could result in continuous process improvement and produce the following outputs:

A. Requested changes and recommended corrective actions

B. Quality checklists and process improvement plan

C. Process improvement plan and quality baseline

D. Quality baseline and project management plan (updates)

95. The project management team monitors specific project results to ensure compliance with pertinent quality standards. Which of the following statements is correct about project quality control?

A. Project quality planning is a key process of project quality control.

B. Project quality control identifies standards that are relevant to the project.

C. Continuous process improvement is part of project quality control.

D. Project quality control requires knowledge of sampling and probability.

96. Your company established an acceptable budget variance of +/– 5% for every project. On a monthly basis, projects may have a budget variance outside the acceptable range, provided this does not happen more than twice in a row. What quality tool should you use to evaluate your project's cost variances over the past two years?

A. Ishikawa diagram

B. Cause and effect diagram

C. Control chart

D. Flowcharting

97. Certain characteristics help identify basic tools of quality. Which list accurately enumerates the key components of a control chart?

A. Major defect, potential causes and effect

B. Upper control limit, x-axis, and lower control limit

C. Activities, decision points, and order of processing

D. Type or category, frequency by cause, and cumulative percentage

98. Your project team was responsible for automatically installing a popular desktop application on 25,000 computers in six countries in Asia, North America, and Europe. Most of the computers have the same configuration, but there might be slight variations in each country and in each line of business within that country. The lead technical architect suggested that you send a technical support representative to verify the installation of every 100th computer. What type of sampling method was recommended?

A. Quota sampling

B. Simple random sampling

C. Stratified sampling

D. Systematic sampling

99. The Project Human Resource Management knowledge area deals with organizing the project team. Which statement is *inaccurate* about the project management team?

A. The project management team is a subset of the project team.

B. The project management team plans, controls, and closes the project.

C. Core team is another name for the project management team.

D. The project management office is a subset of the project management team.

100. Certain constraints may limit the project manager's ability to select project team members. In a trade/labor union environment, which of the following constraints will directly impact the roles and responsibilities that team members may perform in a project?

A. Organizational structure

B. Collective bargaining agreement

C. Economic conditions

D. Interpersonal relationships

101. The project manager may use the following tools and techniques to acquire the project team:

A. Preassignment, negotiation, acquisition, and virtual teams

B. General management skills, training, and co-location

C. Team-building activities, ground rules, and recognition

D. General management skills, training, and rewards

102. You are in the process of forming a project team that includes employees and contractors. A subject matter expert strongly suggested that all of your team members acquire a special technical certification to minimize potential mistakes in the project. What would be the *most* appropriate next step for you to take?

A. Pay for the employees' technical certifications and ask the contractors to pay on their own.

B. Require all employees to earn the technical certification, but not necessarily the contractors.

C. Ask all employees and contractors to earn the technical certification as soon as possible.

D. Prepare a training plan to ensure that all team members have the technical certification.

103. A vendor's response to a request for proposal (RFP) identified three key team members by name who will be working on a project. One team member will perform the initial analysis at the corporate headquarters, and two team members will perform the detailed analysis at various field offices across the country. With regard to acquiring the project team members, the three key team members are considered ______________.

A. Preassigned

B. Negotiated

C. Acquired

D. Virtual

104. There are several ways of acquiring project team members. From the options, which statement *best* describes negotiation as it relates to acquiring project team members?

A. Ask the vendors to list the subject matter experts who will be working in the project in their responses for the request for proposal (RFP).

B. Request functional managers to provide technical specialists with specific skills to work on the project during project initiation.

C. Convince senior managers to hire senior technical consultants and to subcontract parts of the work that are not critical to the schedule.

D. Influence the leadership team to allow some of the team members to work remotely to minimize the travel expenses in the project.

105. The tool or technique used in acquiring project team members by hiring independent consultants or subcontracting portions of the project is called (a) ______________.

A. Preassignment

B. Negotiation

C. Acquisition

D. Virtual team

106. Some of your project team members work in their home offices. Because of the urgency of the project, you have teams that work on three different shifts. Your project is essentially operating ______________.

A. As a 24/7 operation

B. In a fast-track mode

C. Effectively and efficiently

D. As a virtual team

107. Early in the project, you started to identify the training needs of your project team and you established an efficient communication framework to optimize the efficiency and effectiveness of your project team. By performing these actions, you are performing the ______________ process.

A. Human Resource Planning

B. Acquire Project Team

C. Develop Project Team

D. Manage Project Team

108. The Develop Project Team process produces the ______________ output.

A. Team performance assessment

B. Project staff assignments

C. Resource availability

D. Staffing management plan (updates)

109. Developing the project team includes training. The following statements are accurate about training except ______________.

A. Formal training produces the best results in a project environment.

B. Training methods include classroom, mentoring, and coaching.

C. Training requirements may be found in the staffing management plan.

D. Unplanned training may be scheduled as a result of an observation.

110. Which of the following tools or techniques is used in the Develop Project Team process?

A. Preassignment

B. Negotiation

C. Training

D. Acquisition

111. The management team drafted a rewards program for the warehouse redesign project. Having a "model team member of the week" is an example of a ______________ reward.

A. Win-win

B. Win-lose

C. Lose-lose

D. Lose-win

112. You are in the process of drafting a formal rewards program for your project team. The project sponsor indicated a strong dislike for zero sum rewards. Given this, which of the following behaviors should *not* be encouraged?

A. Recognize the team's willingness to work overtime to meet an aggressive schedule.

B. Encourage all team members to submit their weekly progress reports on time.

C. Incorporate a normal distribution when rating the performance of team members.

D. Distribute the performance bonus equally to the team at the end of the project.

113. A senior project manager suggested that you incorporate informal assessments to measure the project team's effectiveness. From the following list, which is an example of an informal performance assessment?

A. MBWA

B. ACWP

C. CWBS

D. CPPC

114. In your organization, the project performance review serves as a critical input to the annual performance review. As a project manager, you are responsible for conducting the project performance review. What can you do to ensure that your project team members will receive a favorable annual performance review?

A. Focus on the positive performance behaviors of all team members

B. Incorporate interim formal reviews throughout the project life cycle

C. Set challenging and aggressive goals for all project team members

D. Coach the team formally and informally whenever opportunities arise

115. In a matrix organization, proper management of the dual reporting relationship can determine the success of a project. In such situations, team members are accountable to the ________________.

A. Functional manager and project manager

B. Project sponsor and project manager

C. Human resource manager and department manager

D. Project leader and end users

116. You are working in a matrix organization. Who is primarily responsible for managing the dual reporting relationship of project team members?

A. Functional manager

B. Project manager

C. Project sponsor

D. Human resource manager

117. Your project uses the principles of 360-degree feedback. Based on these principles, performance evaluations can come from various sources but *not* from ____________.

A. Superiors

B. Peers

C. Subordinates

D. Evaluatees

118. A project team member announced her pregnancy during the regular weekly project meeting. She also indicated that she will be taking a maternity leave at the end of the year. The project manager will need to manage the staffing changes using which process?

A. Perform Integrated Change Control

B. Human Resource Planning

C. Develop Project Team

D. Manage Project Team

119. Which of the following are processes in the Project Communications Management knowledge area?

A. Identify Stakeholders, Plan Communications, Distribute Information, Manage Stakeholder Expectations, and Report Performance

B. Collect Requirements, Plan Communications, Plan Procurements, Administer Procurements, and Scope Verification

C. Collect Requirements, Plan Communications, Distribute Information, Manage Project Team, and Manage Stakeholder Expectations

D. Identify Stakeholders, Distribute Information, Plan Procurements, Administer Procurements, and Performance Reporting

120. The Plan Communications process determines how and when to communicate information to project stakeholders. Which statement below is correct about the Plan Communications process?

A. The Plan Communications process produces the communications management plan as an output.

B. The Plan Communications process requires the communications management plan as an input.

C. The Manage Stakeholder Expectations process needs the communications management plan as an input.

D. The Distribute Information process generates the communications management plan as an output.

121. A new manager who just joined the project was looking for guidelines on how to distribute the monthly progress reports to the steering committee. Such guidelines can be found in the ______________.

A. Project management plan

B. Communications management plan

C. Organizational process assets

D. Performance reports

122. The project manager prepared the weekly progress reports and distributed the same to key project stakeholders. The project manager also responded to an ad hoc request by the finance department to provide an itemized list of capital expenditures. Which process is *best* characterized by these tasks?

A. Report Performance

B. Control Costs

C. Perform Quality Assurance

D. Distribute Information

123. You are managing a government project that impacts three other government agencies. Over time, you developed good professional relationships with the other project managers. In fact, you often find yourself discussing project-related information during your lunch breaks. Such discussions are considered ______________ communication.

A. Internal, vertical

B. Informal, horizontal

C. External, formal

D. Oral, external

124. The following are examples of electronic tools for project management *except* ______________.

A. Web interfaces to project management software

B. Meeting and virtual office support tools

C. Hard-copy document distribution systems

D. Collaborative work management tools

125. Lessons Learned allow current and future projects to avoid repeating the same mistakes. Who is ultimately responsible for conducting Lessons Learned sessions?

A. Project manager

B. Project sponsor

C. Quality manager

D. Project auditor

126. You are working on a major project to upgrade the facilities of a regional beverage distributor. Within the first three months, you realized that it takes a long time to get approvals on purchase orders if the total amount is more than $10,000. Such knowledge should be ______________.

- **A.** Documented as Lessons Learned
- **B.** Noted in the organizational process assets
- **C.** Included in the project records
- **D.** Recorded in the project reports

127. During the execution phase of the project, the project sponsor requested that a monthly progress report be sent to the regulatory compliance department. Which statement *best* describes the proper course of action for the project manager to take?

- **A.** Review the overall project scope to determine if the monthly report is outside the original project scope.
- **B.** Investigate the total effort required to produce the report and only distribute it if no extra effort is required.
- **C.** Modify the communications management plan as needed, as defined in the Perform Integrated Change Control process.
- **D.** Ask the project sponsor to fill out a change request form and submit the same to the change control board.

128. The Report Performance process collects data and distributes performance information to project stakeholders. Which statement below is correct about the tools and techniques that are used in the Report Performance process?

- **A.** The Report Performance process uses variance analysis, forecasting methods, and reporting systems.
- **B.** The most important tools in the Report Performance process are time reporting systems and cost reporting systems.
- **C.** The tools and techniques of the Report Performance process include communication methods and information distribution tools.
- **D.** The Report Performance process utilizes communications requirements analysis, communications technology, and communication models.

129. Which of the following lists include some of the Report Performance process inputs?

- **A.** Work performance information, work performance measurements, and budget forecast
- **B.** Quality control measurements, enterprise environmental factors, and project scope statement
- **C.** Performance measurement baseline, communications management plan, and forecasted completion
- **D.** Approved change requests, organizational process assets, and forecasted completion

130. _______________ is an approved plan for the project work against which project execution is compared and against which deviations are measured for management control.

A. Project management plan

B. Work performance information

C. Quality control measurements

D. Performance measurement baseline

131. A document contains a table with the following columns: WBS element, PV, EV, AC, cost variance, schedule variance, CPI and SPI. The document is *most likely* _______________.

A. A performance report

B. A forecast analysis

C. A risk register

D. An issue log

132. As an information technology project manager, you have used monetary values whenever you perform earned value analysis. Your brother is a civil engineer, and he works mostly on roads and highways. On his projects, it is more important to measure performance based on the length of paved roads and highways instead of costs. Will it be appropriate for him to use earned value analysis on his projects?

A. No, because earned value analysis requires monetary values for planned value, earned value, and actual cost for every WBS element

B. Yes, but he will need to convert the length of paved roads and highways into monetary values for every WBS element

C. Yes, the length of paved roads and highways also can be used for the planned budget, earned value, and actual cost

D. No, because earned value analysis is applicable only to information technology projects and not construction projects

133. Project managers need to manage the expectations of stakeholders. Which communication method is the most effective means of resolving issues with stakeholders?

A. Face-to-face meetings

B. Telephone calls

C. Email

D. Issue tracking logs

134. Which of the following lists contain some of the outputs of the Manage Stakeholder Expectations process?

A. Change requests, forecasts, and communications management plan

B. Change requests, issue log, and project document updates

C. Project management plan updates, change requests, and project document updates

D. Performance reports, change requests, and organizational process assets updates

135. Which knowledge area process attempts both to increase the probability and impact of favorable events and to decrease the likelihood and effect of events that can have adverse consequences on the project?

A. Project Quality Management

B. Project Risk Management

C. Project Procurement Management

D. Project Integration Management

136. In the Project Risk Management knowledge area, the initial risk register is produced during the ______________ process.

A. Plan Risk Management

B. Identify Risks

C. Perform Qualitative Risk Analysis

D. Perform Quantitative Risk Analysis

137. During which Project Risk Management knowledge area processes can the risk register be updated?

A. Plan Risk Management, Identify Risk, and Perform Qualitative Risk Analysis

B. Identify Risk, Perform Qualitative Risk Analysis, and Perform Quantitative Risk Analysis

C. Perform Qualitative Risk Analysis, Perform Quantitative Risk Analysis, Plan Risk Responses, and Monitor and Control Risks

D. Identify Risk, Plan Risk Response, and Monitor and Control Risks

138. Your project used linear values of 0.1, 0.3, 0.5, 0.7, and 0.9 in analyzing the probability and impact of risks. The values correspond to very low, low, moderate, high, and very high, respectively. A project risk with a moderate probability and a high impact will have a risk rating of ______________.

A. 1.20

B. 0.71

C. 0.20

D. 0.35

139. The Identify Risks process uses several tools and techniques. Which information gathering technique would be *most* appropriate if you need to work primarily with subject matter experts to identify the project risks?

A. Brainstorming

B. Delphi technique

C. Interviewing

D. Root cause analysis

140. Your project used linear values of 0.1, 0.3, 0.5, 0.7, and 0.9 in analyzing the probability and impact of risks. From the following list of descriptions, which will *most likely* be assigned to the values respectively?

A. Moderate, neutral, average, high, and avoid

B. Low, medium, high, average, and very low

C. Very low, low, moderate, high, and very high

D. Very high, high, moderate, low, and very low

141. Using decision tree analysis, you calculated that Project A has an expected monetary value (EMV) of $65,000. In contrast, Project B has an EMV of $75,000. Which project is a better option?

A. Project A, because it has a lower EMV

B. Project B, because it has a higher EMV

C. Neither project, because the difference in EMV is not significant

D. Both, in order to maximize the results

142. Given a false decision node with a cost of $–50M with scenario probabilities of A (55%, $100M) and B (45%, $25M), what is the EMV for this decision node?

A. 0.00

B. 16.25

C. 66.25

D. 75.00

143. Given a false decision node with a cost of $–50M with scenario probabilities of A (55%, $100M) and B (45%, $25M), what are the net path values for A and B, respectively?

A. $27.50M and $–11.25M

B. $50.00M and $–25.00M

C. $55.00M and $11.25M

D. $72.50M and $2.50M

144. Which of the following risk response strategies are appropriate for negative risks or threats?

A. Avoid, transfer, and/or mitigate

B. Exploit, share, and/or enhance

C. Avoid, transfer, and/or recognize

D. Share, enhance, and/or recognize

145. Which risk response strategies are appropriate for positive risks or opportunities?

A. Avoid, transfer, and/or mitigate

B. Exploit, share, and/or enhance

C. Avoid, transfer, and/or recognize

D. Share, enhance, and/or recognize

146. Which risk response strategy can be applied to both negative risks (threats) and positive risks (opportunities)?

A. Mitigate

B. Exploit

C. Share

D. Accept

147. Which of the following statements *best* characterizes residual risks and secondary risks?

A. Secondary risks arise if the risk response generated a residual risk that is much larger than the original risk that was identified in the risk register.

B. Secondary risks remain after the implementation of planned responses, whereas residual risks arise as a direct outcome of implementing a risk response.

C. Residual risks remain after the implementation of planned responses, whereas secondary risks arise as a direct outcome of implementing a risk response.

D. Risk response strategies for negative risks or threats generate residual risks, whereas positive risks or opportunities result in secondary risks.

148. In the middle of your project, you noticed that some risks positively and negatively impacted your budget and schedule contingency. What do you need to perform to determine if you have adequate contingency to cover the remaining project risks?

A. Risk reassessment

B. Risk audits

C. Variance and trend analysis

D. Reserve analysis

149. The following lists contain outputs of the Monitor and Control Risks process *except* for ________.

A. Risk register updates and change requests

B. Project management plan updates and change requests

C. Organizational process assets updates and project management plan updates

D. Risk register updates and risk-related contract decisions

150. The previous project manager left you a spreadsheet that summarized uncertain events with the following columns: probability, impact, priority, and ownership. You are *most likely* looking at what type of document?

A. Risk register

B. Requested changes

C. Recommended corrective actions

D. Recommended preventive actions

151. Which Project Procurement Management knowledge area process produces the procurement documents?

A. Conduct Procurements

B. Plan Procurements

C. Close Procurements

D. Administer Procurements

152. Your project issued a purchase order for computer hardware to be delivered at the end of the month. The unit cost is based on the manufacturer's suggested retail price. The purchase order includes an incentive for early delivery. This is an example of a ________ contract.

A. Fixed-price

B. Cost-reimbursable

C. Time and material

D. Cost-plus-incentive-fee

153. Your company hired an offshore company to develop a hardware prototype for a new product line. The project contract states that the seller will be paid based on approved expenses. A fee will be paid depending on the total expenses. What type of contract did the project use?

A. CPF

B. CPFF

C. CPIF

D. T&M

154. Your health care project requires a very complex service. Technical capability, management approach, financial capacity, references and proprietary rights are examples of ______________.

A. Procurement documents

B. Source selection criteria

C. Procurement statement of work

D. Procurement document package

155. The Project Procurement Management knowledge area process deals with purchasing or acquiring products, services, or results from outside the project team. Which statement is correct about obtaining responses to bids and proposals?

A. The Plan Procurements process produces the procurement management plan.

B. The Conduct Procurements process generates the selected sellers and the procurement contract award.

C. The Conduct Procurements process uses bidder conferences and proposal evaluation techniques.

D. The Administer Procurements process uses a contract change control system and claims administration.

156. Which statement is correct about the tools and techniques that are used in the Conduct Procurements process?

A. The qualified sellers list contains all sellers that the buyer asked to submit a response to a request for proposal or a request for quotation.

B. The Conduct Procurements process uses the following tools and techniques: bidder conferences, proposal evaluation techniques, and procurement negotiations.

C. The buyer prepares the procurement document package and formally invites the sellers to prepare a bid for the requested products or services.

D. The seller prepares a proposal in response to procurement document package; the proposal constitutes a formal and legally binding offer.

157. Bidder conferences normally are held with prospective sellers prior to the preparation of bids or proposals. Other names for bidder conferences include all of the following *except* ______________.

A. Contractor conferences

B. Vendor conferences

C. Pre-bid conferences

D. Virtual conferences

158. Which Project Procurement Management knowledge area process uses the following tools and techniques: independent estimates, advertising, and procurement negotiations?

A. Plan Procurements

B. Conduct Procurements

C. Administer Procurements

D. Close Procurements

159. Which documents are prepared by the sellers and highlight their ability and willingness to deliver a product, service, or result for the buyer?

A. Agreements

B. Memos of understanding

C. Proposals

D. Letters of intent

160. You are in the process of negotiating with potential suppliers of plastic raw materials for your project. You separated the criteria into technical, economic, experience, and reputation categories. At this stage, what process are you performing?

A. Plan Procurements

B. Conduct Procurements

C. Activity Definition

D. Cost Estimating

161. When selecting sellers, the source selection criteria may include all of the following *except* ________________.

A. Samples of the sellers' products

B. References from previous customers

C. Credentials of the management team

D. Copies of contracts from other clients

162. Your organization drafted a procurement document package in anticipation of upcoming legislation that will drastically impact how you perform your day-to-day operations. If you distribute the procurement document package to selected sellers, what can you expect in return?

A. Proposals

B. Invoices

C. Quotations

D. Requirements

163. The project sponsor suggested that the selection committee evaluate sellers based solely on quantitative criteria. Which criterion should be *excluded* in the evaluation process?

- **A.** Assign a numerical weight to each requirement
- **B.** Perform forced ranking of all sellers on each criterion
- **C.** Rate each seller based on the size of previous projects
- **D.** Solicit expert opinions from key industry leaders

164. Several publishers provided different estimates for your project's training manuals. If the range of estimates is too wide, what should you do?

- **A.** Select the seller with the lowest estimate
- **B.** Award the contract to the highest bidder
- **C.** Perform your own independent estimate
- **D.** Pick the seller that you are familiar with

165. Assume that you are the leader of the proposal evaluation team. You asked your team members to come up with ideas on how to evaluate sellers. From the following list, which evaluation technique would *most likely* require further refinements?

- **A.** Include both subjective and objective evaluation criteria
- **B.** Rate and score the proposals based on multiple categories
- **C.** Assess the proposal based on the evaluator's narrative summary
- **D.** Assign predefined weightings to business and user requirements

166. A contract obligates the seller to provide the services to the buyer, and it requires the buyer to pay the seller upon receipt of the services. A contract remains valid unless ________________.

- **A.** The buyer cannot pay the seller because of unforeseen financial difficulties
- **B.** The contract contains clauses that will violate existing laws and regulations
- **C.** The seller does not have the financial resources to deliver the service
- **D.** One party's legal department cancels the contract with sufficient notice

167. What is the main purpose of the Administer Procurements process?

- **A.** To commit the seller to deliver the services as agreed
- **B.** To make sure that the buyer lists the project requirements
- **C.** To get a better price from sellers of the same products
- **D.** To ensure both parties meet their contractual obligations

168. You received a contract from a seller to provide human resources consulting services for your project. There is a clause in the contract that can potentially impact your organization. Your project could be delayed if you ask the legal department to review the contract. What should you do?

A. Sign the contract because you have already been authorized to do so in the project charter

B. Ask for a delay in the project schedule so that the legal department can review the contract

C. Cross out the contract clause in question, initial it, and then sign the contract to avoid delays

D. Seek legal advice on what to do with the contract and assess the impact to the schedule

169. Which statement is true about the procurement management plan?

A. The procurement management plan is a subsidiary of the project management plan.

B. The procurement management plan is similar to the project management plan.

C. The procurement management plan is a subsidiary of the communications management plan.

D. The procurement management plan is an output of the contract administration process.

170. The procurement management plan, part of the project management plan, is used as an input by the ______________.

A. Administer Procurements and Close Procurements processes

B. Conduct Procurements and Close Procurements processes

C. Conduct Procurements, Administer Procurements, and Close Procurements processes

D. Plan Procurements, Conduct Procurements, and Close Procurements processes

171. You are working on a small web development project. Who would normally be responsible for paying the seller in accordance with the terms in the contract?

A. The company's accounts payable system

B. The senior leadership team

C. The functional owner of the project

D. The sponsor of the web development project

172. Within the Project Procurement Management knowledge area process, procurement documentation is an output of the _______________ process and an input to the _______________ process.

A. Plan Procurements; Conduct Procurements

B. Plan Procurements; Administer Procurements

C. Conduct Procurements; Administer Procurements

D. Administer Procurements; Close Procurements

173. How is the Close Procurements process different from the Close Project or Phase process?

A. The Close Procurements process supports the Close Project or Phase process.

B. Only the Close Project or Phase process ensures that one party is not at default.

C. The Close Procurements process is exclusively applied at a project's end.

D. The Close Procurements process is not the responsibility of the project manager.

174. The prototype tests conducted for your project by a seller failed. You have no choice but to terminate the contract early. What should you do next?

A. Mutually terminate the contract early and close the project

B. Consider the seller in default because of the failed prototype tests

C. Pay the seller an early termination fee as stated in the contract

D. Review the contract's terms and conditions for early termination

175. You closed a contract on an early phase of the project despite some unresolved claims with the seller. You reasoned that the seller will rectify the unresolved claims on the next project phase. If you still cannot agree on how to resolve the issues after the next phase, what should you do?

A. Convince the seller to go to arbitration

B. Threaten the seller with possible legal actions

C. Withhold all payments to the seller immediately

D. Review the contract's terms and conditions

176. Disagreements may arise between the seller and the buyer. If both parties accepted and signed a contract, which outcome is *least* desirable in case of disagreements?

A. Mutually resolve the disagreements

B. Ask the buyer to pay the penalty

C. Convince the seller that it is wrong

D. Take the issue to binding arbitration

177. The Close Procurements process uses several inputs. Which list contains the correct set of inputs?

A. Project management plan and procurement documentation

B. Enterprise environmental factors, project scope statement, and work breakdown structure

C. Procurement management plan, contract statement of work, and project management plan

D. Contract, contract management plan, selected sellers, and performance reports

178. Which statement is correct about the tools and techniques used in Project Procurement Management?

A. The Plan Procurements process uses standard forms and expert judgment.

B. The Close Procurements process uses procurement audits, negotiated settlements, and records management systems.

C. The Plan Procurements process uses bidder conferences and records management systems.

D. The Conduct Procurements process uses weighting systems and independent estimates.

179. Procurement audits are structured reviews that recognize good practices and identify questionable items during the preparation and administration of procurement contracts. Which Project Procurement Management process uses procurement audits as a tool or a technique?

A. Plan Procurements process

B. Conduct Procurements process

C. Administer Procurements process

D. Close Procurements process

180. As part of the Close Procurements process, certain organizational assets may be updated. Which of the following lists identifies the correct portions of the organizational assets that may be updated during Contract Closure?

A. Procurement file, deliverable acceptance, and Lessons Learned documentation

B. Correspondence, payment schedules and requests, and procurement file

C. Seller performance evaluation documentation and deliverable acceptance

D. Payment schedules and requests, procurement file, and correspondence

181. The Administer Procurements process was completed for your project. What output should be produced in the next Project Procurement Management process?

A. Closed procurements and organizational process assets updates

B. Procurement documentation and project management plan updates

C. Select sellers and procurement management plan updates

D. Qualified sellers list and procurement document package

182. Which two Project Procurement Management knowledge area processes update the organizational process assets?

A. Plan Procurements and Conduct Procurements

B. Conduct Procurements and Administer Procurements

C. Plan Procurements and Administer Procurements

D. Administer Procurements and Close Procurements

183. A colleague who works in a nonproject management capacity at your organization volunteered to serve on a committee of a local chapter even though she is not a member of the Project Management Institute (PMI®). During her term, she was accused of questionable behaviors. Does the *Code of Ethics and Professional Conduct* apply to her?

A. No, because she is not involved in the project management profession

B. No, because she is not a member of the PMI®

C. Yes, because she is involved as a volunteer in a local PMI® chapter

D. Yes, because she is still expected to act professionally outside work

184. You are still working as a project manager, but you let your PMI® membership and certification expire. Does the *Code of Ethics and Professional Conduct* still apply to you?

A. Yes, because you are still actively practicing as a project manager

B. No, because you are not a PMI® member and you do not hold any PMI® certifications

C. Yes, because you once had a valid project management certification

D. No, because you are no longer a member of the PMI®

185. The *Code of Ethics and Professional Conduct* does *not* apply to which personal situation?

A. Project management practitioner, non-PMI® member and not a holder of PMI® credentials

B. PMI® credential holder but not practicing project management and a non-PMI® member

C. Occasional volunteer at PMI® events but not a project management practitioner

D. Submitted application for a PMI® credential but not a member of a local PMI® chapter

186. You are working on a public safety project for a local government. On an extremely rare occasion, one of your project deliverables may violate a privacy law, but you know, in good faith, that it will not necessarily cause any personal harm. What should you do?

A. Don't ask and don't tell because it may only happen on extremely rare occasions

B. There is no need to bring it up because it will not cause any personal harm

C. Disclose only when asked, so as not to delay the project and cause unnecessary panic

D. Inform the project management team about the scenario and address it accordingly

187. You are very excited to work on a large, multimillion dollar project. Shortly after taking on the role, you realized that you do not have the experience to deliver it successfully. If you remove yourself from the project, it could be the end of your career with the company. You really want this project to be successful, and you are willing to do anything to make it work, but you do not want to look incompetent. What should you do?

A. Schedule special training to get the necessary skills

B. Ask a more senior project manager to mentor you

C. Inform key stakeholders of the gap in your experience

D. Work extra hours to make up for your deficiencies

188. You overheard a colleague who recently earned a PMI® credential discussing specific difficult questions on the exam. She was talking to another colleague who earned the PMI® credential years ago. What should you do?

A. Politely remind your colleagues to refrain from discussing specific exam questions

B. Ignore the conversation—you should not have been listening in the first place

C. Report the incident to the PMI® Ethics Review Committee as soon as possible

D. Do not do anything—nobody will benefit from their conversations anyway

189. You have always had suspicions about the professional relationship of another project manager with a major supplier. You strongly feel that other project managers share the same suspicions regarding this person. What is the *most appropriate* course of action in this situation?

A. Ask other project managers if they share the same suspicions

B. Report the suspicions to the company's ethics review committee

C. Start gathering facts to substantiate your personal suspicions

D. Stop being envious of the other project manager's relationship

190. During one of your weekly meetings, one of your team members shared a very funny but racial joke. The other person for whom the joke was intended did not mind it, and he started laughing as well. As a project manager, you are expected to ______________.

A. Laugh with the team to foster team development

B. Ask for similar jokes to relieve the stress of the project team

C. Tell the team member not to do it again, outside the meeting

D. Remind the team, at the meeting, that such jokes are not tolerated

191. You serve on the board of directors of a local PMI® chapter. The chapter was voting on a motion that could potentially benefit your employer but not you personally. As part of your duty as a project management practitioner, you should ______________.

A. Support the board's motion and highlight the benefits of hiring your employer

B. Abstain from the discussions to avoid the perception of conflict of interest

C. Point out the deficiencies on all proposals, including your employer's proposal

D. Assess all options and then offer your impartial and expert opinion to the group

192. You realized that you have *not* been fair to one of your project team members. What would be an appropriate course of action?

A. Apologize to the other project team member

B. Remind yourself to be fair next time

C. Do not do anything, as it was a one-time event

D. Take corrective actions as appropriate

193. You were hired to assess a software integration project. Based on your review, it became obvious that the project should be cancelled. If you make such recommendation, however, your contract will be terminated as well. You really need the money, and you cannot afford to lose this contract. You should ______________.

A. Tell the truth even if it means losing your contract

B. Make suggestions on how to correct the problems

C. Prolong the contract until you can find another one

D. Ask the project sponsor to redefine the overall scope

194. The *Code of Ethics and Professional Conduct* discusses aspirational and mandatory standards. From the following list, which one is an aspirational standard of fairness?

A. Proactively disclose any potential conflict of interest

B. Demonstrate transparency when making decisions

C. Not discriminate against others based on race or gender

D. Apply rules of the organization without prejudice

195. From the following list, which action is an example of a mandatory standard of honesty?

A. Make commitments and promises in good faith

B. Make others feel safe to tell the truth

C. Provide accurate information in a timely manner

D. Do not engage in dishonest behaviors

196. A project management contractor did not lie in her presentation but withheld some information to convince the buyer to purchase her services. Such behavior demonstrates failure to meet the __________ of the *Code of Ethics and Professional Conduct.*

A. Aspirational standard of honesty

B. Mandatory standard of responsibility

C. Mandatory standard of honesty

D. Aspirational standard of fairness

197. Which of the following statements is *not* acceptable with regard to aspirational and mandatory standards of honesty?

A. Provide accurate information in a timely manner

B. Sometimes make implied commitments in good faith

C. Remain truthful in all project communications

D. Avoid dishonest behavior for personal gain

198. A project manager makes it a point to remind his foreign workers that they can be deported if they do not finish their deliverables on time. Can this behavior be considered abusive?

A. No, because it meets the aspirational and mandatory standards of honesty

B. Yes, because it creates an intense feeling of fear on the part of the foreign workers

C. No, because his duty of loyalty requires him to give the timely reminders

D. Yes, because foreign workers place little importance on project deadlines

199. What is the *best* way to resolve issues that may arise when there is a potential clash between duty of loyalty and conflict of interest?

A. Adhere to the duty of loyalty

B. Disclose the conflict entirely

C. Avoid the conflict of interest

D. Do not engage in the discussion

200. A person's responsibility, legal or moral, to promote the best interest of an organization or other person with whom he or she is affiliated is called ______________.

A. Duty of loyalty

B. Social obligation

C. Membership affinity

D. Professional responsibility

201. A(n) ______________ is a collection of projects, programs, and other portfolios.

A. Program management office

B. Program

C. Portfolio

D. Organizational process asset

202. In balanced and strong matrix structures, the project manager's titles are ______________ and ______________, respectively.

A. Project coordinator, project leader

B. Project leader, project expeditor

C. Project expeditor, project manager

D. Project manager, project manager

203. Which statement is correct with regard to process names and project management process groups?

A. Develop Project Charter is a process within the Planning process group.

B. Develop Project Management Plan is a process within the Initiating process group.

C. Direct and Manage Project Execution is a process within the Monitoring and Controlling process group.

D. Develop Project Charter is a process within the Initiating process group.

204. Which of the following processes is *not* part of the Planning process group?

A. Verify Scope

B. Collect Requirements

C. Define Scope

D. Create WBS

205. Which statement accurately describes interviews and focus groups?

A. Interviews are one-on-one; focus groups are cross-functional.

B. Interviews are one-one-one; focus groups are usually facilitated by a moderator.

C. Interviews are cross-functional; focus groups are usually facilitated by a moderator.

D. Interviews are usually facilitated by a moderator; focus groups are cross-functional.

Chapter 4

Answers to Practice Test A

Answer Key for Practice Test A

1. A.
2. D.
3. B.
4. C.
5. D.
6. A.
7. C.
8. D.
9. A.
10. B.
11. A.
12. D.
13. A.
14. B.
15. C.
16. D.
17. B.
18. A.
19. A.
20. A.
21. A.
22. A.
23. C.
24. C.
25. A.
26. C.
27. C.
28. C.
29. D.
30. A.
31. B.
32. A.
33. D.
34. C.
35. A.
36. B.
37. D.
38. C.
39. A.
40. B.
41. D.
42. D.
43. A.
44. B.
45. C.
46. A.
47. B.
48. C.
49. B.
50. C.
51. A.
52. D.
53. A.
54. B.
55. B.
56. C.
57. C.
58. A.
59. A.
60. A.
61. D.
62. C.
63. B.
64. D.
65. A.
66. A.
67. B.
68. C.
69. B.
70. D.
71. C.
72. B.
73. A.
74. B.
75. C.
76. C.
77. A.
78. A.
79. A.
80. D.
81. C.
82. D.
83. A.
84. D.
85. A.
86. D.
87. A.
88. A.
89. D.
90. B.
91. D.
92. C.
93. A.
94. A.
95. D.
96. C.
97. B.
98. D.
99. D.
100. B.
101. A.
102. D.
103. A.
104. B.
105. C.
106. D.
107. C.
108. A.
109. A.
110. C.
111. B.
112. C.
113. A.
114. D.
115. A.
116. B.
117. D.
118. A.
119. A.
120. A.
121. B.
122. D.
123. B.
124. C.
125. A.
126. A.
127. C.
128. A.
129. A.
130. D.
131. A.
132. C.
133. A.
134. C.
135. B.
136. B.
137. C.
138. D.
139. B.
140. C.
141. B.
142. B.
143. B.
144. A.
145. B.
146. D.
147. C.
148. D.
149. D.
150. A.
151. B.
152. A.
153. A.
154. B.
155. C.
156. B.
157. D.
158. B.
159. C.
160. B.
161. D.
162. A.
163. D.
164. C.
165. C.
166. B.
167. D.
168. D.
169. A.
170. C.
171. A.
172. D.
173. A.
174. D.
175. D.
176. D.
177. A.
178. B.
179. D.
180. A.
181. A.
182. D.
183. C.
184. B.
185. A.
186. D.
187. C.
188. C.
189. C.
190. D.
191. B.
192. D.
193. A.
194. B.
195. D.
196. C.
197. B.
198. B.
199. B.
200. A.
201. C.
202. D.
203. D.
204. A.
205. B.

1. A. Within the context of a project, the Project Integration Management knowledge area unifies and consolidates the various project management process groups (Initiating, Planning, Executing, Monitoring and Controlling, and Closing) from the beginning until the end of the project.

   The Project Scope Management knowledge area entails identifying all of the work required in order to bring the project to a successful completion. The Project Quality Management knowledge area determines and defines the quality policies, objectives, activities, and responsibilities within the project. The Project Human Resource Management knowledge area deals with the resourcing, organization, and management of the project team.

   Reference: *PMBOK® Guide, 4th Ed.*, page 71.

2. D. Experienced project managers apply various project management tools and techniques depending on the context of the project. The PMBOK® Guide specifically states that "the perception that a particular process is not required does not mean that it should not be addressed." Therefore, it is the project manager's responsibility to seek professional expertise as needed, as well as to adjust the degree of rigor of various processes and procedures depending on the context of the project.

   Reference: *PMBOK® Guide, 4th Ed.*, page 72

3. B. The project charter serves as the formal authorization for a project or a project phase. The Perform Integrated Change Control process includes reviewing, approving, and managing changes in the project. The project management plan captures the actions required to consolidate and coordinate various supplementary plans. Directing and managing project execution occurs after the initiation and planning of the project.

   Reference: *PMBOK® Guide*, 4th Ed., page 71

4. C. The development of the project charter occurs during project initiation, whereas the development of the project management plan happens during project planning and the performance of integrated change control takes place during project monitoring and controlling. Managing stakeholder expectations is part of the Project Communications Management knowledge area.

   Reference: *PMBOK® Guide*, 4th Ed., pages 43, 71

5. D. The outputs of the Integrated Change Control process include change request status updates, project management plan updates, and project document updates. The outputs of the Close Project or Phase process include organizational process assets updates.

   Reference: *PMBOK® Guide*, 4th Ed., page 73

6. A. Expert judgment is used in all of the Project Integration Management knowledge area processes. The project management information system is used in the Direct and Manage Project Execution process. Change control meetings belong to the Perform Integrated Change Control process. Facilitated workshops are part of the Project Scope Management knowledge area.

Reference: *PMBOK® Guide*, 4th Ed., page 73

7. C. After the development of the project management plan, the project manager may proceed with the Direct and Manage Project Execution process. The Project Integration Management knowledge area processes flow as follows: Develop Project Charter, Develop Project Management plan, Direct and Manage Project Execution, Monitor and Control Project Work, Perform Integrated Change Control, and Close Project or Phase.

Reference: *PMBOK® Guide*, 4th Ed., page 71

8. D. Inputs of the Close Project or Phase process consist of the project management plan, accepted deliverables, and organizational process assets. The expert judgment technique mentioned in the question is common to all six processes in the Project Integration Management knowledge area process; therefore, you need to be intimately familiar with the inputs, tools and techniques, and outputs of each process in order to correctly answer a question similar to this one on the exam.

Reference: *PMBOK® Guide*, 4th Ed., page 73

9. A. The project initiator or sponsor will either create the project charter or delegate that duty to the project manager. The sponsor's or initiator's signature on the charter authorizes the project. The project manager may assist in developing the project charter, but the project initiator or sponsor is the one who can authorize it.

Reference: *PMBOK® Guide*, 4th Ed., page 74

10. B. As a result of internal business needs or external influences, an organization may initiate some sort of business case, needs assessment, or plan prior to the development of the project charter. The project scope statement and the project management plan are developed after the project charter. Likewise, detailed cost estimates and work breakdown structure occur later in the project management process.

Reference: *PMBOK® Guide*, 4th Ed., pages 74–75

11. A. The project charter normally includes direct and indirect references to the project purpose or justification, project objectives and success criteria, high-level requirements and project description, high-level risks, a summary milestone schedule, and a summary budget. The rest of the options are associated with the development of the project scope statement and not the project charter.

Reference: *PMBOK® Guide*, 4th Ed., pages 77–78

12. D. For an external customer, a project manager will need a contract (when applicable) in order to prepare the project charter. The project scope statement is part of the Planning process, of which the key output is the project management plan. Work performance information is an input to the Perform Integrated Change Control process.

Reference: *PMBOK® Guide*, 4th Ed., page 76

13. A. The SOW summarizes the products or services that the project will deliver. Key components of the SOW include the business need, product scope description, and strategic plan. The other options are enterprise environmental factors, which makes them incorrect. Take note, however, that similar to the SOW, those enterprise environmental factors serve as inputs in the development of the project charter.

Reference: *PMBOK® Guide*, 4th Ed., page 75

14. B. Examples of enterprise environmental factors include government standards, infrastructure, and project management information systems. Business need, product scope description, and strategic plan make up the project statement of work (SOW), which makes Option A incorrect. Although not necessarily incorrect, standardized cost estimating data and industry risk study information are examples of commercial databases. Likewise, automated tool suite and configuration management are examples of project management information systems.

Reference: *PMBOK® Guide*, 4th Ed., pages 14, 76

15. C. Financial control procedures entail reporting time, reviewing expenditures, reviewing disbursements, and assigning accounting codes using standard provisions for contracts. Examples of project closure guidelines include auditing the final project and validating the product. Risk control procedures aim to categorize and assess risks using information about probabilities and impact. Organizational standard processes impact the entire organization through standards, policies, checklists, and so on.

Reference: *PMBOK® Guide*, 4th Ed., pages 32–33, 76.

16. D. The two categories of organizational process assets are organizational processes and procedures, and corporate knowledge base. The other options are incorrect because they are mainly examples of organizational processes and procedures.

Reference: *PMBOK® Guide*, 4th Ed., pages 32–33

17. B. An organizational corporate knowledge base includes the process measurement database, project files, historical information, the Lessons Learned knowledge base, the issue and defect management database, the configuration management knowledge base, and the financial database. The other options are incorrect because they include organizational processes and procedures—not organizational corporate knowledge base.

Reference: *PMBOK® Guide*, 4th Ed., page 33

18. A. The SOW is a narrative description of products or services to be delivered by the project. A document that formally initiates the project and identifies the project manager describes the project charter. Option C describes the procurement statement of work, and Option D is the project scope statement.

Reference: *PMBOK® Guide*, 4th Ed., page 75

19. A. The process used to develop the project charter is used to define and document the high-level requirements and risks of the project. The same process also deals with measurable project objectives, related success criteria, summary milestone schedule, summary budget, and project approval requirements. The other options are incorrect.

Reference: *PMBOK® Guide*, 4th Ed., pages 77–78

20. A. The correct answer is the project charter. Some of the items included in the project charter are the high-level requirements, high-level risks, measurable project objectives, related success criteria, summary milestone schedule, summary budget, and project approval requirements.

In contrast, the project management plan deals with the remaining process groups that follow the initiating and planning of the project: executing, monitoring and controlling, and closing. The project scope statement contains, among other things, the detailed product scope description, project deliverables, project constraints, and assumptions. The project statement of work is an input in the development of the project charter.

Reference: *PMBOK® Guide*, 4th Ed., pages 77–78

21. A. The project initiator or sponsor may delegate the creation of the project charter to the project manager, and thus must provide this information. Although the other options are inputs in the development of the project charter, the project initiator or sponsor is the best answer.

Reference: *PMBOK® Guide*, 4th Ed., page 74

22. A. The contract, project statement of work, enterprise environmental factors, and organizational process are inputs to the development of the project charter. Although the other options are highly similar, they are incorrect. The project scope statement, requirements documentation, and approved changed requests are part of scope planning. Project management processes are not an input in the development of the project charter. The project management plan, work performance information, and rejected change request are inputs to the Monitor and Control Project Work process.

Reference: *PMBOK® Guide*, 4th Ed., pages 75–77

23. C. The project management plan encompasses all the outputs of the Planning Process Group. Given this, the selected project management processes can be found in the project management plan. The other options are incorrect.

Reference: *PMBOK® Guide*, 4th Ed., page 81

24. C. The Direct and Manage Project Execution process, as the name implies, deals with the actual execution of the project management plan by the project manager and the project team. They will act to produce the project deliverables, as outlined in the project scope statement. The other processes are incorrect because they are parts of the Monitoring and Controlling Process Group.

Reference: *PMBOK® Guide*, 4th Ed., page 83

25. A. Projects may produce or deliver tangible and intangible deliverables during project execution. From the options provided, staff skills (through training or skill transfer) are the only intangible deliverable. Highway lighting and traffic signs are tangible deliverables.

Reference: *PMBOK® Guide*, 4th Ed., page 91

26. C. Collecting the project data and facilitating the forecasting of cost and schedule are actions that are associated with the Direct and Manage Project Execution process, which is part of the Executing process group.

Reference: *PMBOK® Guide*, 4th Ed., page 83

27. C. Take note that change requests include corrective action, preventive action, defect repair, and updates, thus making Option C the correct answer. Approved corrective actions are documented, authorized directions that the project manager and the project team must perform in order to bring the project back on the path defined in the project management plan. In contrast, approved preventive actions help reduce the likelihood of risks that can contribute to an unfavorable project outcome. Approved change requests deal with the expansion and contraction of project scope. Approved defect repairs aim to correct product defects that were found during the quality inspection or the audit process.

Reference: *PMBOK® Guide*, 4th Ed., page 87

28. C. Approved change requests deal with the expansion and contraction of project scope, among other approved changes. Approved corrective actions are documented, authorized directions that the project manager and the project team must perform in order to bring the project back on the path defined in the project management plan. In contrast, approved preventive actions help reduce the likelihood of risks that can contribute to an unfavorable project outcome. Approved defect repairs aim to correct product defects that were found during the quality inspection or the audit process.

Reference: *PMBOK® Guide*, 4th Ed., page 86

29. D. Expert judgment defines a process that guides a project team executing the project management plan. The project management information system and work performance information system are not processes—they are tools. Change requests that were neither approved nor accepted to be included in the project, along with their supporting information, are collectively called rejected change requests. Expert judgment is provided by the project manager and the project management team using specific knowledge and skills, and it aids the project team in executing various project activities outlined in the project management plan.

Reference: *PMBOK® Guide*, 4th Ed., page 86

30. A. Because the initiating and planning phases already were completed, the project manager is likely to be at the execution phase of the project. One process during project execution is Direct and Management Project Execution, which produces the following outputs: deliverables, work performance information, project management plan updates, project document updates, and change requests, which include corrective and preventive actions, and defect repair. The other options are incorrect because they list outputs for the Monitoring and Controlling, and Closing processes of a project.

Reference: *PMBOK® Guide*, 4th Ed., pages 87–88

31. B. As defined in the project management plan, work performance information is routinely collected to determine which deliverables have been delivered, to record costs that have been incurred, and to gain an understanding of the extent of the remaining tasks. Changes in the project scope, policies, and other factors are normally documented as change requests. Deliverables are unique and tangible products, results, or capabilities that complete the project upon delivery. Implemented change requests prove that change requests were implemented, implemented corrective actions correct future project performance, and implemented preventive actions help reduce the negative consequences of project risks.

Reference: *PMBOK® Guide*, 4th Ed., page 87

32. A. The Monitor and Control Project Work process is performed throughout the project, during the project's initiating, planning, executing, and closing processes.

Reference: *PMBOK® Guide*, 4th Ed., page 89

33. D. Although comparing actual performance to the project management plan, assessing project performance, and monitoring the implementation of approved changes and managing project risks are important components of project monitoring and control, they do not necessarily relate to the question. To facilitate earned value management techniques, the project manager must establish project baselines from which overall progress and variances can be calculated and compared.

Reference: *PMBOK® Guide*, 4th Ed., page 89

34. C. Cost performance index (CPI) = earned value (EV) / actual cost (AC)

Research Project A = $500 / $505 = 0.99

Research Project B = $455 / $450 = 1.01

Research Project C = 1.08

Research Project D = 0.94

A CPI less than 1.0 indicates a cost overrun—the project is spending more than the planned estimates. In contrast, a CPI greater than 1.0 signifies a cost underrun of the estimate. Therefore, Research Project C is the most cost efficient. Essentially, the project receives $1.08 in value of actual completed work for every dollar that it spends.

Reference: *PMBOK® Guide*, 4th Ed., page 181

35. A. Cost variance (CV) = earned value (EV) – actual cost (AC)

CV = $747 – $691 = $56

Reference: *PMBOK® Guide*, 4th Ed., page 181

36. B. Important processes of configuration management are configuration identification, configuration status accounting, and configuration verification and auditing. Architecture configuration and control is vague relative to the other options.

Reference: *PMBOK® Guide*, 4th Ed., page 95

37. D. Documented change requests must be either accepted or rejected by the change control board. If change requests are documented, it is implied that they must be reviewed. Deferring a change request is essentially a rejection.

Reference: *PMBOK® Guide*, 4th Ed., page 94

38. C. The project management plan, which contains the project scope statement as part of the scope baseline, may need to be updated as a result of an approved change request. Closing the project may require updates to organizational process assets. Note that the administrative closure procedure and contract closure procedure are inputs of the Close Project or Phase process and not the Perform Integrated Change Control process.

Reference: *PMBOK® Guide*, 4th Ed., pages 98–99

39. A. The Perform Integrated Change Control process tools and techniques include change control meetings and expert judgment. A change control board can be used within the change control meetings. The earned value technique helps assess the overall health of the project as part of the Monitoring and Controlling processes—not the Perform Integrated Change Control process.

Reference: *PMBOK® Guide*, 4th Ed., page 98

40. B. The Close Project or Phase process must be performed at the end of each phase, especially on a multiphase project, and at the end of the overall project. Such closure includes finalizing a specific portion of the project scope and all of its associated activities. By doing so, Lessons Learned from a phase can then be applied to the succeeding phase. If the Close Project or Phase process is performed only at the end of the project, then the project team will not be able to formally take advantage of Lessons Learned during project execution.

Reference: *PMBOK® Guide*, 4th Ed., pages 99–100

41. D. The organizational process assets updates involve updating project files, project or phase closure documents, and historical information. In contrast, the close procurements procedure entails providing the seller with formal written notice that the contract is now completed. Within the context of this question, record retention and performance review are not formally recognized terms and are incorrect.

Reference: *PMBOK® Guide*, 4th Ed., pages 102, 344

42. D. One of the closing procedures entails collecting and updating project files, project or phase closure documents, and historical information. This occurs as part of the Close Project or Phase process, which also transitions the final product, service, or result, although that is not the case in this question. Although it might be appropriate to ask the sponsor to reconsider, a specific mandate is required for you to perform this closing procedure. Settling any contractual claim is a procedure that is carried out as part of the Administer Procurements process, and making alterations and retesting is not a valid option.

Reference: *PMBOK® Guide*, 4th Ed., page 102

43. A. Inputs to the Close Project or Phase process include the project management plan, accepted deliverables, and organizational process assets. Enterprise environmental factors and organizational process assets are both inputs to the Develop Project Charter, Develop Project Management Plan, and Close Project or Phase processes. Contract (when applicable), project charter, and project scope statement are not inputs of the Close Project or Phase process.

Reference: *PMBOK® Guide*, 4th Ed., pages 73, 101

44. B. Expert judgment is a tool and technique used by all Project Integration Management knowledge area processes.

Reference: *PMBOK® Guide*, 4th Ed., page 73

45. C. Organizational process assets that can influence the Close Project or Phase process include, but are not limited to, project or phase closure guidelines or requirements (for example, project audits, project evaluations, and transition criteria), along with the historical information and Lessons Learned knowledge base (for example, project records and documents, all project closure information, information regarding previous project selection decisions, previous project performance information, and information from the risk management effort). The other options are incorrect.

Reference: *PMBOK® Guide*, 4th Ed., page 101

46. A. The Close Procurements process lists the step-by-step actions involved in formally closing all contracts. It is important to briefly revisit the Close Procurements process to determine if the collection agency's claims have any merit. Although the project is no longer your responsibility, it would be unprofessional to simply terminate the telephone call without having a plan of action.

Reference: *PMBOK® Guide*, 4th Ed., page 341

47. B. Organizational process assets include project files, project closure documents, and historical information. The project manager updates these documents as part of the Close Project or Phase process. The configuration management system is a collection of formal documented procedures for maintaining the functional and physical characteristics of artifacts associated with a project. Work performance information is a collection of routine activity updates as the project progresses. Enterprise environmental factors include internal and external stimuli that can influence the success of the project.

Reference: *PMBOK® Guide*, 4th Ed., page 102

48. C. As an output of the Close Project or Phase process, project closure documents indicate the formal completion of the project and transfer of project deliverables to the operations group. If the project is canceled prior to its completion, the same document will indicate the reason for the cancellation and the state of incomplete activities.

Official acceptance of project deliverables is documented in the formal acceptance documentation. Historical information includes Lessons Learned that can be used by future projects. Although the Close Project or Phase process may be conducted in multiphase projects—producing the project closure documents as an output—the end of a project phase does not necessarily mean the end of a project.

Reference: *PMBOK® Guide*, 4th Ed., page 102

49. B. The creation of the WBS is part of the Project Scope Management knowledge area process called Create WBS.

Reference: *PMBOK® Guide*, 4th Ed., page 116

50. C. The Project Scope Management knowledge area processes include the following: Collect Requirements, Define Scope, Create WBS, Verify Scope, and Control Scope. A detailed description of the project is created and defined during the Define Scope process, producing the project scope statement, and the Create WBS process subdivides the major project deliverables into smaller components. The other options are incorrect.

Reference: *PMBOK® Guide*, 4th Ed., page 103

51. A. The Define Scope process uses the following tools and techniques: product analysis, alternatives identification, facilitated workshops, and expert judgment. Tools and techniques for the Collect Requirements process include interviews, focus groups, questionnaires, and surveys. The Create WBS process entails use of decomposition. The Control Scope process uses variance analysis as a tool and technique.

Reference: *PMBOK® Guide*, 4th Ed., page 104

52. D. The group creativity tool and technique uses brainstorming, the Delphi technique, idea/mind mapping, and several other tools and techniques to be used in group activities. Observation, also called *job shadowing*, entails directly viewing individuals in their environment and observing how they perform their jobs or tasks. Facilitated workshops use focused sessions to define the product requirements by bringing key cross-functional stakeholders together. Group decision-making techniques use several methods to reach group decisions—unanimity, majority, plurality, and dictatorship.

Reference: *PMBOK® Guide*, 4th Ed., page 107–109

53. A. The WBS is a hierarchical decomposition of the work to be performed on a project. Because the second level of the WBS listed the major parts of a bicycle, it would be reasonable to assume that the succeeding level will further break down each major part into smaller parts (front wheel and rear wheel).

Reference: *PMBOK® Guide*, 4th Ed., page 116

54. B. Decomposition is the only tool and technique in the Create WBS process; inspection is for the Verify Scope process and variance analysis is for the Control Scope process. Within the Project Scope Management knowledge area, facilitated workshops is one of the tools and techniques of the Collect Requirements and Define Scope processes.

Reference: *PMBOK® Guide*, 4th Ed., pages 114, 116

55. B. The work breakdown structure (WBS) is a hierarchical decomposition of the work to be performed on a project. In contrast, an organizational breakdown structure (OBS) hierarchically depicts the project organization in order to properly relate the work packages to the performing organizational units. As their names imply, the risk breakdown structure and resource breakdown structure deal with project risks and resources, respectively.

Reference: *PMBOK® Guide*, 4th Ed., page 220

56. C. Formal acceptance of the completed project scope and associated deliverables from the stakeholders occurs during the Verify Scope process. Sequentially, the Collect Requirements process occurs first, followed by the Define Scope process. Outputs of the Define Scope process serve as inputs to the Create WBS process, which is then followed by the Verify Scope process. The Control Scope process is the last process within Project Scope Management.

Reference: *PMBOK® Guide*, 4th Ed., pages 113, 123

57. C. Inspection entails measuring, examining, and verifying that the project deliverables adhere to the requirements. Other industries may adopt similar but different techniques such as reviews, walk-throughs, or audits. Product analysis is a tool or technique for the Verify Scope process, decomposition is for the Create WBS process, and variance analysis is for the Control Scope process.

Reference: *PMBOK® Guide*, 4th Ed., page 124

58. A. The Control Scope process uses the following inputs: project management plan, work performance information, requirements documentation, and organizational process assets. The tool and technique for this process is variance analysis. Through the use of this tool and technique, the Control Scope process produces the following outputs: work performance measurements, organizational process assets, change requests, project management plan updates, and project document updates.

Reference: *PMBOK® Guide*, 4th Ed., pages 125–128

59. A. Project performance measurements are used to assess the magnitude of variation. Therefore, the remaining options are incorrect.

Reference: *PMBOK® Guide*, 4th Ed., page 127

60. A. An analysis of project's progress against the project scope can result in issuance of change requests to the scope baseline or components of the project management plan. The configuration management system is a collection of formal documented procedures for maintaining the functional and physical characteristics of artifacts associated with a project. The change control system is a subset of the configuration management system. The requirements traceability matrix is a table that connects requirements to their origins and traces them throughout the lifetime of the project.

Reference: *PMBOK® Guide*, 4th Ed., pages 111, 125

61. D. Outputs of the Control Scope process include updates to the following: work performance measurements, organizational process assets, change requests, project management plan updates, and project document updates.

Reference: *PMBOK® Guide*, 4th Ed., page 128

62. C. As a result of performing the Create WBS process, the scope baseline will be created. As part of the Control Scope process, the scope baseline may need to be updated (as part of project management plan updates) if an approved change request will have an impact on the scope. Scope baseline is not an output of Collect Requirements, Define Scope, and Verify Scope processes.

Reference: *PMBOK® Guide*, 4th Ed., pages 122, 128

63. B. The Sequence Activity process uses the PDM. The other options are incorrect.

Reference: *PMBOK® Guide*, 4th Ed., page 138

64. D. Rolling wave planning is a form of progressive elaboration planning where near-term work is planned in more detail and future work is planned at a higher level. As more information becomes known about upcoming events, it can then be decomposed into activities. In contrast, WBS branches with sufficient definition should be decomposed to the work package level.

Reference: *PMBOK® Guide*, 4th Ed., page 135

65. A. The PDM uses nodes (boxes or rectangles) to represent activities; arrows connect the nodes to show the dependencies. Leads and lags allow for acceleration and delays of successor activities, respectively. The three types of dependency determination used to define the sequence of activities include mandatory (hard logic), discretionary (soft logic), and external dependencies.

Reference: *PMBOK® Guide*, 4th Ed., pages 138–141

66. A. Finish-to-start is the most commonly used type of precedence relationship in the PDM. The other options are incorrect.

Reference: *PMBOK® Guide*, 4th Ed., page 138

67. B. The Estimate Activity Resources process is closely coordinated with the Estimate Costs process. Inputs of the Estimate Activity Resources process include enterprise environmental factors, organizational process assets, activity list, activity attributes, and resource calendars. Outputs of the Estimate Activity Resources process include activity resource requirements, resource breakdown structure, and project document updates.

Reference: *PMBOK® Guide*, 4th Ed., pages 138–141

68. C. Draw a network diagram based on the table. The critical path includes tasks A, D, E, H, and J. The sum of durations of each of these tasks is 24 days.

Reference: *PMBOK® Guide*, 4th Ed., page 146

69. B. $E = (O + 4M + P) / 6 = (8 + 4 \times 14 + 16) / 6 = 13.33$

Reference: *PMBOK® Guide*, 4th Ed., page 150

70. D. Draw a network diagram based on the table. The critical path includes tasks A, D, E, H, and J. The sum of the durations of each task in the critical path is 24 days.

Reference: *PMBOK® Guide*, 4th Ed., page 154

71. C. Graphical representations of the project schedule consist of project schedule network diagrams, bar charts, and milestone charts. The project schedule is too broad relative to the correct answer. Milestone charts, the correct answer, show the start or completion dates of major deliverables and key external interfaces. Project schedule networks diagram are suited for showing task dependencies, whereas bar charts are ideal for showing the duration of the tasks.

Reference: *PMBOK® Guide*, 4th Ed., page 157

72. B. Outputs of the Develop Schedule process include project schedule, schedule data, schedule baseline, and project document updates, which may include activity resource requirements (updates), activity attributes (updates), calendar (updates), and risk register (updates). The other options are incorrect because the lists contain outputs from the Control Schedule process.

Reference: *PMBOK® Guide*, 4th Ed., pages 157–160

73. A. The Control Schedule process allows the project manager to assess the current status of the project schedule, influence the factors that create schedule changes, ascertain changes in the project schedule, and manage actual changes. The rest of the statements are incorrect.

Reference: *PMBOK® Guide*, 4th Ed., page 160

74. B. The Control Schedule process is performed as part of the Monitoring and Controlling process group. The other options are incorrect.

Reference: *PMBOK® Guide*, 4th Ed., pages 43, 160

75. C. The scenario lists the outputs of the Control Schedule process. Therefore, the following inputs are required: organizational process assets, project documents, and project management plan. The other options are incorrect because they list inputs to the Develop Schedule process.

Reference: *PMBOK® Guide*, 4th Ed., pages 152, 160

76. C. Schedule performance index (SPI) = earned value (EV) / planned value (PV) = 0.80

In a nutshell, the project is 80 percent on schedule, meaning the project is 20 percent behind. Therefore, corrective actions most likely will be required to bring the project back on track. The other options are incorrect.

Reference: *PMBOK® Guide*, 4th Ed., page 183

77. A. An SPI higher than 1.0 indicates that the project is getting more value (earned value) than the budgeted cost for the work scheduled to be completed at that point in time (planned value). If the earned value and planned value are indeed accurate, then there is nothing to worry about in terms of schedule progress.

Reference: *PMBOK® Guide*, 4th Ed., page 189

78. A. Schedule variance (SV) = earned value (EV) – planned value (PV) = 888 – 999 = –111

Reference: *PMBOK® Guide*, 4th Ed., page 189

79. A. The Project Cost Management knowledge area processes are Estimate Costs, Determine Budget and Control Costs.

Reference: *PMBOK® Guide*, 4th Ed., page 165

80. D. The range of the ROM will be wide during the initiation phase of a project—particularly if there is no historical data from which to base the estimates. Given this, the range of –50 to +50 percent appears to be reasonable.

Reference: *PMBOK® Guide*, 4th Ed., page 168

81. C. Parametric estimating relies on historical data to come up with estimates, plus a specific value of the existing project. Depending on the estimating model, the estimates can be accurate. It is likely that the construction estimator came up with the cost estimate based on square footage and material (for example, carpet, tiles, or hardwood).

Analogous cost estimating derives estimates based on comparable projects. Such a perception of similarity must be based on facts and not just appearance; otherwise, the estimates will not be accurate. Bottom-up estimates are based on estimates of each lower-level activity. The estimates are then rolled up to come up with the estimates. Vendor bid analysis is based on comparing the estimates of multiple vendors.

Reference: *PMBOK® Guide*, 4th Ed., pages 171–173

82. D. The accuracy and reliability of analogous or parametric estimating is dependent on the correctness of the historical information and the scalability of the model (for example, applicability to small, medium, or large projects). It is also imperative that parameters are based on quantifiable factors.

Reference: *PMBOK® Guide*, 4th Ed., pages 177–178

83. A. After performing the Control Costs process, the following outputs may be updated: budget forecasts, organizational process assets, project management plan, and project documents. The remaining options are incorrect because they are inputs, not outputs, of the Control Costs process.

Reference: *PMBOK® Guide*, 4th Ed., pages 187–188

84. D. The ETC based on atypical variances = budget at completion (BAC) – cumulative earned value to date (EV^C) = 500 – 400 = 100

CPI^C is needed only if the ETC needs to be calculated based on typical variances.

Reference: *PMBOK® Guide*, 4th Ed., pages 184–185

85. A. The Plan Quality process uses the COQ technique.

Reference: *PMBOK® Guide*, 4th Ed., page 195

86. D. The PMBOK® Guide considers the quality management approaches espoused by Deming, Juran, Crosby, and others as proprietary. In contrast, nonproprietary management approaches include Total Quality Management (TQM), Six Sigma, Failure Mode and Effect Analysis, Voice of the Customer, and other approaches not associated with a single individual or expert.

Reference: *PMBOK® Guide*, 4th Ed., page 190

87. A. Failure costs are also known as the cost of poor quality. Internal failures and external failures are the two types of failure costs. The cost of quality is a superset of failure costs.

Reference: *PMBOK® Guide*, 4th Ed., page 195

88. A. Quality assurance deals with the process of auditing the quality requirements and the results to ensure appropriate standards, whereas quality control focuses on the monitoring and recording results of executing the quality activities to assess performance. The other options are incorrect.

Reference: *PMBOK® Guide*, 4th Ed., pages 189, 201

89. D. The process improvement plan includes analysis of process boundaries, process configuration, process metrics, and targets for improved performance in order to identify and remove non–value-added activities in a process. Configuration management is not part of a process improvement plan.

Reference: *PMBOK® Guide*, 4th Ed., page 201

90. B. Project Quality Management knowledge area processes include the following: Plan Quality, Perform Quality Assurance, and Perform Quality Control. Quality planning tools and techniques, quality audits, and process analyses are used in the Perform Quality Assurance process to convert inputs into outputs.

Reference: *PMBOK® Guide*, 4th Ed., page 202

91. D. Change requests are involved in the Perform Quality Assurance and Perform Quality Control processes and, as such, need to follow the Perform Integrated Change Control process before they can be approved and implemented. Any undocumented changes should not be processed or implemented.

Reference: *PMBOK® Guide*, 4th Ed., pages 191, 205

92. C. Project quality audits may be scheduled or random. Given that the project quality audit was scheduled to occur at least once a month, the project manager needs to contact the project quality auditor to conduct the audit as planned.

Reference: *PMBOK® Guide*, 4th Ed., page 204

93. A. A quality audit is an assessment performed by an unbiased individual or entity to determine if project activities conform to the organization and project policies and procedures. A configuration management system is a collection of formal documented procedures for maintaining the functional and physical characteristics of artifacts associated with a project. A change control system is a subset of the configuration management system. Process analysis identifies processes for improvement as defined in the process improvement plan.

Reference: *PMBOK® Guide*, 4th Ed., page 204

94. A. The quality activities stated in the question characterize project quality assurance. Performing quality assurance in a project produces the following outputs: organizational process assets updates, change requests (recommended corrective actions and/or preventive actions), project management plan updates, and project document updates. The other options are outputs of quality planning, not quality assurance; hence, they are incorrect answers.

Reference: *PMBOK® Guide*, 4th Ed., pages 201, 205

95. D. Project quality control requires knowledge of statistical quality control to determine the causes of unsatisfactory results. The other options are incorrect.

Reference: *PMBOK® Guide*, 4th Ed., page 206

96. C. An Ishikawa diagram is another name for a cause and effect diagram, so both options can be eliminated as correct answers. Flowcharting helps to analyze how problems occur. Given this, control chart is the correct answer because it can be used to determine if a process is stable—in this particular case, the monthly project budget.

Reference: *PMBOK® Guide*, 4th Ed., page 209

97. B. The following basic tools of quality have these corresponding components:

- Cause and effect diagram—major defect, potential causes and effect
- Control chart—upper control limit, x-axis, and lower control limit
- Flowcharting—activities, decision points, and order of processing
- Pareto chart—type or category, frequency by cause, and cumulative percentage

Reference: *PMBOK® Guide*, 4th Ed., pages 208, 213

98. D. Systematic sampling selects every kth element within the sampling frame. In this particular example, the technical support representative will check every 100th computer within the sampling frame of 25,000 computers.

Reference: *PMBOK® Guide*, 4th Ed., page 212

99. D. The project management team (also called core, executive, or leadership team) plans, controls, and closes the project. The project management team is a subset of the project team. The project management team is a subset of the project management office.

Reference: *PMBOK® Guide*, 4th Ed., page 215

100. B. In a unionized organization, the collective bargaining agreement will directly impact the roles and responsibilities that team members may perform in a project. The collective bargaining agreement is basically a labor contract between an employer and one or more unions that defines the parties' relationships in a workplace environment.

Reference: *PMBOK® Guide*, 4th Ed., page 225

101. A. The Acquire Project Team process uses the following tools and techniques: preassignment, negotiation, acquisition, and virtual teams. The other options are incorrect because they are tools and techniques used in the Develop Project Team process.

Reference: *PMBOK® Guide*, 4th Ed., pages 227–228

102. D. Although the other options are acceptable, the most correct answer—partly because of its generalized nature—is to prepare a training plan to make sure that all team members have the technical certification. The plan can also include ways to help team members obtain certifications that would benefit the project. The other less correct options potentially can be included as part of the training plan.

Reference: *PMBOK® Guide*, 4th Ed., page 225

103. A. The three key team members are considered preassigned because their assignments to the project are known in advance. Staff negotiation entails getting the project team members from various functional or departmental areas. Acquired staff means hiring consultants or subcontracting the work. The three key members will be working as a virtual team, but the fact that they were known in advance makes the preassigned option a more correct answer than the virtual option.

Reference: *PMBOK® Guide*, 4th Ed., page 227

104. B. Preassignment involves knowing the project team members prior to the start of the project. Hiring consultants or subcontracting the work is considered an acquisition. Virtual teams work remotely. Although all of these may require some form of negotiation, the correct answer is requesting the functional managers (or similar roles) to provide the staff with the proper skills to work in a particular time frame.

Reference: *PMBOK® Guide*, 4th Ed., pages 227–228

105. C. Preassignment means knowing project team members in advance. Staff negotiation entails getting the project team members from various functional or departmental areas. Virtual teams, although working on the same project, perform the work in different locations. Hiring independent consultants or subcontracting portions of the project are examples of acquisition. Acquisition is the correct answer.

Reference: *PMBOK® Guide*, 4th Ed., page 228

106. D. The project team is basically working around the clock and in a fast-track mode. We know that they are working 24 hours per day (8 hours × 3 shifts = 24 hours); however, it is not known if they work on weekends as well. Thus, 24×7 is incorrect. We also do not know if the project is running effectively and efficiently. Given this, virtual team is the best answer because home offices and shift work are characteristics of a virtual team.

Reference: *PMBOK® Guide*, 4th Ed., page 228

107. C. Enhancing the competencies of team members and facilitating the interactions among them, especially when performed early in the project, enhance the overall project team performance. Such actions are part of the Develop Project Team process.

Reference: *PMBOK® Guide*, 4th Ed., pages 217, 229

108. A. Team performance assessment and enterprise environmental factors updates are outputs of the Develop Project Team process. The other options are incorrect because they are outputs of the Acquire Project Team process.

Reference: *PMBOK® Guide*, 4th Ed., pages 235–236

109. A. Although training can be formal or informal, it is inaccurate to claim that formal training produces the best results in a project environment. In some cases, informal training may be comparable to or even better than formal training.

Reference: *PMBOK® Guide*, 4th Ed., pages 217, 232

110. C. Training is one of the tools and techniques for the Develop Project Team process. Preassignment, negotiation, and acquisition are tools and techniques of the Acquire Project Team process.

Reference: *PMBOK® Guide*, 4th Ed., pages 217, 232

111. B. A win-lose (zero sum) reward limits the recognition of desirable behaviors to one or a limited number of project team members. Such reward programs may even foster unhealthy competition among team members. It is desirable to encourage win-win rewards program instead so that all team members have the same opportunity to earn the reward without excluding others.

Reference: *PMBOK® Guide*, 4th Ed., page 234

112. C. Examples of win-lose (zero sum) rewards include team member of the month—only certain members can earn it. Using normal distribution when rating the performance of team members means that some members will be rated as "low" and some as "high," with the rest left in the middle (average). Such reward systems can foster unhealthy competition among team members. The PMBOK® Guide encourages rewarding win-win situations instead—that is, giving all team members the same opportunity to earn the rewards. For example, rewards could be given for submitting progress reports on time, putting in extra hours to meet a deadline, and achieving other performance goals.

Reference: *PMBOK® Guide*, 4th Ed., page 234

113. A. Actual cost of work performed (ACWP, contract work breakdown structure (CWBS), and cost plus percentage of cost (CPPC) are not informal performance assessments. Only management by walking around (MBWA) is an example of an informal assessment.

Reference: *PMBOK® Guide*, 4th Ed., page 235

114. D. All of the options are essentially correct, but coaching the team formally and informally throughout the project will help keep team members focused on achieving individual and collective project goals. Good work can be recognized and rewarded, and poor work can be adjusted as needed.

It is good to focus on positive performance, but negative performance should not be hidden or ignored. Formal reviews throughout the project are needed and should be complemented with informal reviews. In addition to being challenging and aggressive, goals must be achievable and realistic.

Reference: *PMBOK® Guide*, 4th Ed., page 235

115. A. In a matrix organization, team members are accountable to a functional manager and a project manager. The other options are incorrect.

Reference: *PMBOK® Guide*, 4th Ed., pages 28–32

116. B. In a matrix organization, the project manager is primarily responsible for managing the dual reporting relationship of project team members. The other options are incorrect.

Reference: *PMBOK® Guide*, 4th Ed., pages 28–32

117. D. Based on the 360-degree feedback principles, performance evaluations can be received from all sources that interact with the person being evaluated, such as superiors, peers, and subordinates, but not from the evaluatees themselves.

Reference: *PMBOK® Guide*, 4th Ed., page 238

118. A. Change requests such as staffing changes are handled through the Perform Integrated Change Control process. The other options are incorrect.

Reference: *PMBOK® Guide*, 4th Ed., page 242

119. A. Project Communications Management knowledge area processes include Identify Stakeholders, Plan Communications, Distribute Information, Manage Stakeholder Expectations, and Report Performance.

Reference: *PMBOK® Guide*, 4th Ed., page 243

120. A. The Plan Communications process produces the communications management plan as an output. The Manage Stakeholder Expectations process needs the communications management plan as an input (as a subsidiary component of the project management plan), but the question is asking about the Plan Communications process. The Distribute Information process does not generate the communications management plan as an output.

Reference: *PMBOK® Guide*, 4th Ed., pages 244, 259, and 263

121. B. The communications management plan, a subsidiary of the overall project management plan, contains the guidelines on how to distribute information within the project. Organizational process assets are too broad and performance reports are too narrow (too specific) to be correct answers.

Reference: *PMBOK® Guide*, 4th Ed., pages 256–257

122. D. The Distribute Information process entails gathering and disseminating relevant project information as outlined in the communications management plan. The Distribute Information process also includes ad hoc or unplanned requests for information.

The Report Performance process is also a plausible answer, but there was no mention of baseline data or resource usage relative to the achievement of project objectives. The Control Costs process is also very tempting because of references to the finance department and capital expenditures. Weekly progress reports are generic in nature, however, and are usually not specific regarding project costs or budgets. The Perform Quality Assurance process is incorrect.

Reference: *PMBOK® Guide*, 4th Ed., page 258

123. B. We know that the communication as described is oral informal (ad hoc conversation) and horizontal (with peers). Although the other project managers work for other government agencies, it is not clear if the professional relationships can be considered internal or external. We know that the communication is not vertical and not formal. Therefore, the correct answer is informal, horizontal.

Reference: *PMBOK® Guide*, 4th Ed., page 245

124. C. Hard copy is not an electronic tool.

Reference: *PMBOK® Guide*, 4th Ed., page 260

125. A. Project managers have a professional obligation to conduct Lessons Learned sessions for all projects with key internal and external stakeholders, particularly if the project yielded less than desirable results.

Reference: *PMBOK® Guide*, 4th Ed., page 261

126. A. The purchase order approval process obviously merits a review. Although improvement of the process may be outside the project scope, such knowledge should be documented as Lessons Learned. The other options are incorrect because organizational process assets include Lessons Learned documentation, project records, project reports, and other documentation. Lessons Learned are considered parts of project records. Project reports may include Lessons Learned.

Reference: *PMBOK® Guide*, 4th Ed., page 261

127. C. The best answer is to modify the communications management plan as needed, as defined in the Perform Integrated Change Control process. The Perform Integrated Change Control process defines how to handle change requests, including changes that may impact the project scope and effort. Depending on the magnitude of the changes, a change request may not be needed.

Reference: *PMBOK® Guide*, 4th Ed., pages 260, 271

128. A. The Report Performance process uses the following tools and techniques: variance analysis, forecasting methods, communication methods, and reporting systems. Communication methods and information distribution tools are tools and techniques of the Distribute Information process. In contrast, communications requirements analysis, communications technology, and communication models are tools and techniques of the Plan Communications process.

Reference: *PMBOK® Guide*, 4th Ed., pages 268–270

129. A. Inputs of the Report Performance process include work performance information, work performance measurements, budget forecast, project management plan updates, and organizational process assets.

Reference: *PMBOK® Guide*, 4th Ed., pages 267–268

130. D. Performance measurement baseline is an approved plan for the project work against which the project execution is compared, and deviations are measured for management control. The project management plan includes performance measurement baseline information. The other options are incorrect.

Reference: *PMBOK® Guide*, 4th Ed., page 267

131. A. Based on the columns, it appears that the project is using earned value analysis to determine the performance of the project. The document could very well be a forecast, but the lack of estimate at completion and estimate to complete make performance report a better answer. The columns indicate that the document is not a risk register or an issue log.

Reference: *PMBOK® Guide*, 4th Ed., pages 270–271

132. C. Monetary value (dollar amount) is the most common unit of measure in earned value analysis. It is also possible, however, to use other units of measure such as labor hours, square footage, and cubic meters of concrete for calculation purposes.

Reference: *PMBOK® Guide*, 4th Ed., pages 270–271

133. A. If distance is not an issue and availability is not an issue, face-to-face meetings are the most effective means of resolving issues with stakeholders.

Reference: *PMBOK® Guide*, 4th Ed., pages 261–262

134. C. Outputs of the Manage Stakeholder Expectations process include organizational process assets updates, change requests, project management plan updates, and project document updates.

Reference: *PMBOK® Guide*, 4th Ed., page 265

135. B. The Project Risk Management knowledge area process deals with increasing the probability and impact of favorable project events as well as decreasing the probability and impact of adverse project events.

Reference: *PMBOK® Guide*, 4th Ed., page 273

136. B. The Identify Risks process produces the risk register as an output.

Reference: *PMBOK® Guide*, 4th Ed., pages 274, 288

137. C. The risk register is initially created as an output of the Identify Risk process. The risk register is then updated as part of the following processes: Perform Qualitative Risk Analysis, Perform Quantitative Risk Analysis, Plan Risk Responses, and Monitor and Control Risks.

Reference: *PMBOK® Guide*, 4th Ed., page 274

138. D. Risk rating = probability × impact = 0.5 × 0.7 = 0.35

Reference: *PMBOK® Guide*, 4th Ed., pages 279, 281

139. B. Brainstorming sessions generate ideas on potential project risks. Interviewing and root cause analysis can augment the list of project risks that were identified as part of the brainstorming process. Only the Delphi technique, however, helps achieve consensus when dealing with subject matter experts.

Reference: *PMBOK® Guide*, 4th Ed., pages 286–287

140. C. In ascending order, 0.1 usually represents very low probability/impact compared to 0.9, which normally equates to very high probability/impact. Therefore, the correct descriptive labels for the values are very low, low, moderate, high, and very high, respectively. The other options are incorrect because the descriptions are not in the correct ascending order.

Reference: *PMBOK® Guide*, 4th Ed., pages 291–292

141. B. With all other things being equal, a project with a higher EMV is more favorable than a project with a lower EMV. EMV, a statistical concept used in quantitative analysis, is used to analyze potential outcomes based on uncertain conditions.

Reference: *PMBOK® Guide*, 4th Ed., page 298

142. B. Calculate the net path value for A and B.

A = –50 + 100 = \$50M

B = –50 + 25 = \$–25M

Multiply the net path value by the probability of each scenario.

A = 50 × 55% = \$27.5M

B = –25 × 45% = \$–11.25M

The EMV for the false decision node is 27.5 + –11.25 = \$16.25M.

Reference: *PMBOK® Guide*, 4th Ed., pages 298–299

143. B. Calculate the net path value for A and B.

A = –50 + 100 = \$50M

B = –50 + 25 = \$–25M

The probabilities of 55 percent and 45 percent, respectively, are not relevant in this question.

Reference: *PMBOK® Guide*, 4th Ed., pages 298–299.

144. A. Risk response strategies for negative risks or threats include avoid, transfer, mitigate, and/or accept. Risk response strategies for positive risks or opportunities include exploit, share, enhance, and/or accept. Recognize is not a risk response strategy.

Reference: *PMBOK® Guide*, 4th Ed., pages 303–305

145. B. Risk response strategies for negative risks or threats include avoid, transfer, mitigate, and/or accept. Risk response strategies for positive risks or opportunities include exploit, share, enhance, and/or accept. Recognize is not a risk response strategy.

Reference: *PMBOK® Guide*, 4th Ed., pages 303–305

146. D. Risk response strategies for negative risks or threats include avoid, transfer, and/or mitigate. Risk response strategies for positive risks or opportunities comprise exploit, share, and/or enhance. If a risk cannot be reduced or eliminated, acceptance can be adopted as a risk response strategy. Acceptance can be applied to both threats and opportunities.

Reference: *PMBOK® Guide*, 4th Ed., pages 263–264

147. C. Residual risks remain after the implementation of planned responses, whereas secondary risks arise as a direct outcome of implementing a risk response.

Reference: *PMBOK® Guide*, 4th Ed., pages 305–306

148. D. Reserve analysis compares the remaining contingency reserve in the budget and/or schedule to determine if the remaining amount can adequately cover the remaining risks in the project.

Reference: *PMBOK® Guide*, 4th Ed., page 311

149. D. Risk-related contract decisions are outputs of the Plan Risk Responses process.

Reference: *PMBOK® Guide*, 4th Ed., pages 311–312

150. A. A risk register normally contains descriptions of project risks along with probability, impact, priority, and ownership. Status (open, closed, and so forth), dates (open date, trigger date, etc.) and notes may also be included in the risk register. Probability, impact, priority, and ownership are not likely to be part of the other options.

Reference: *PMBOK® Guide*, 4th Ed., pages 311–312

151. B. The Plan Procurements process produces the procurement documents along with the procurement management plan, procurement statements of work, make-or-buy decisions, source selection criteria, and change requests.

Reference: *PMBOK® Guide*, 4th Ed., page 325

152. A. The situation above is not a time and material contract. Cost-reimbursable is a superset of cost-plus-incentive-fee, so it can be eliminated as a choice. The situation as described can very well be a fixed-price (lump-sum) or a cost-plus-incentive-fee contract; however, the latter is often associated with purchases when actual work needs to be performed (for example, paving a road, computer programming, and so forth) instead of simply shipping a regular product. Therefore, fixed-price is considered correct.

Reference: *PMBOK® Guide*, 4th Ed., pages 322–323

153. A. The situation above is not a time and material (T&M) contract. The project used a cost-plus fee (CPF) because the seller will be reimbursed for its expenses and will be paid a fee based on a portion of the total expenses. Cost-plus-fixed fee (CPFF) means that the seller will be reimbursed for its expenses (costs) plus a fixed fee regardless of costs. In contrast, cost-plus incentive fee (CPIF) pays the seller a predetermined incentive based on the achievement of certain milestones.

Reference: *PMBOK® Guide*, 4th Ed., pages 324–325

154. B. Within the context of the Plan Procurements process, source selection criteria incorporate several factors, including the seller's technical capability, management approach, financial capacity, and references. Other source selection criteria may also consider the ownership of intellectual property rights and proprietary rights.

Reference: *PMBOK® Guide*, 4th Ed., pages 327–328

155. C. All of the options are correct; however, the question asks about obtaining responses to bids and proposals. Option C describes actions that relate to requesting seller responses.

Reference: *PMBOK® Guide*, 4th Ed., pages 331–333

156. B. All of the statements are correct; however, only one option lists some of the tools and techniques that are used in the Conduct Procurements process—bidder conferences, proposal evaluation techniques, independent estimates, expert judgment, advertising, Internet search, and procurement negotiations.

Reference: *PMBOK® Guide*, 4th Ed., pages 331–333

157. D. Other names for bidder conferences include contractor conferences, vendor conferences, and pre-bid conferences.

Reference: *PMBOK® Guide*, 4th Ed., page 331.

158. B. The Conduct Procurements process uses the following tools and techniques: bidder conferences, proposal evaluation techniques, independent estimates, expert judgment, advertising, Internet search, and procurement negotiations.

Reference: *PMBOK® Guide*, 4th Ed., pages 331–333

159. C. Prepared by the sellers, proposals highlight their ability and willingness to supply a product, service, or result to the buyer. The proposal constitutes a contract if accepted by the buyer.

Reference: *PMBOK® Guide*, 4th Ed., page 330

160. B. The Conduct Procurements process includes deciding which vendor or supplier the project will use. The process considers several factors, including price, technical approach, and management style, in the negotiation process. The Plan Procurements process is incorrect because it deals with what to purchase, when, and how—not the negotiation process itself. The Activity Definition and Cost Estimating processes are incorrect.

Reference: *PMBOK® Guide*, 4th Ed., page 332

161. D. When selecting sellers, the buyer may consider several factors to properly evaluate the sellers' ability to supply the products, services, or results that the project needs. Such evaluation criteria may include samples of the sellers' products, references from previous customers, and credentials of the management team. Copies of contracts from other clients may also be included in the evaluation criteria; however, because of privacy laws and confidentiality agreements, the seller may not be able to disclose the contracts from other customers.

Reference: *PMBOK® Guide*, 4th Ed., pages 327–328

162. A. Sellers prepare proposals in response to a procurement document package. The proposal will then be used to evaluate and select one or more bidders (sellers).

Reference: *PMBOK® Guide*, 4th Ed., page 330

163. D. All the options can be quantified (numerical weight, forced ranking, and seller ratings) except for soliciting opinions from key industry leaders.

Reference: *PMBOK® Guide*, 4th Ed., page 331

164. C. In the case of widely varying estimates from multiple suppliers, you should perform your own independent estimate. An independent estimate (should-cost estimate) enables the buyer to assess the validity of the seller's estimate. Low price may mean that the seller did not understand the requirements. In contrast, high price may indicate that the seller is proposing a higher-quality product or service.

Reference: *PMBOK® Guide*, 4th Ed., page 332

165. C. It is acceptable to include both subjective and objective evaluation criteria when selecting sellers. Objective criteria such as ratings, scores, and weights generally are preferred to subjective criteria. Given this, the evaluator's narrative summary is less preferred relative to the other options.

Reference: *PMBOK® Guide*, 4th Ed., pages 331–332

166. B. A signed contract is mutually binding, and any modifications must be accepted by both parties. A contract or portions of it may become invalid if it contains clauses or provisions that will violate existing laws and regulations.

Reference: *PMBOK® Guide*, 4th Ed., page 333

167. D. The Administer Procurements process ensures that both parties, the buyer and the seller, meet their contractual obligations. It is imperative that the buyer stipulates the project requirements in the contract and that the seller delivers the services as agreed upon. Getting a better price is a component of selecting the sellers and not administering the contracts.

Reference: *PMBOK® Guide*, 4th Ed., page 335

168. D. Although contractual terms are met within a project, contracts that can have organizational impacts should be reviewed by legal professionals. The project charter authorizes the project manager to expend resources to deliver the project. The project manager, however, may not be in a position to enter into a binding contract with other parties. It is also premature to ask for delays in the project schedule without first getting input from legal professionals. Crossing out a contract clause can also be problematic without proper legal advice.

Reference: *PMBOK® Guide*, 4th Ed., page 335

169. A. The procurement management plan is a subsidiary of the project management plan. it is produced as an output of the Plan Procurements process. The other options are incorrect.

Reference: *PMBOK® Guide*, 4th Ed., pages 324–325

170. C. The Conduct Procurements, Administer Procurements, and Close Procurements processes all use the procurement management plan, part of the project management plan, as an input. Option C is more correct than just having the Administer Procurements and Close Procurements or Conduct Procurements and Close Procurements processes by themselves.

Reference: *PMBOK® Guide*, 4th Ed., pages 314, 330, 337, and 343

171. A. On smaller projects, the company's accounts payable system would normally pay the seller. The project management team reviews and approves the payment to ensure adherence to the terms of the contract.

Reference: *PMBOK® Guide*, 4th Ed., page 339

172. D. Within the Project Procurement Management knowledge area processes, procurement documentation is an output of the Administer Procurements process and an input to the Close Procurements process.

Reference: *PMBOK® Guide*, 4th Ed., pages 340, 343

173. A. At the end of the project or at the end of each major phase of a large project, the Close Procurements process supports the Close Project or Phase process by ensuring delivery of final deliverables within the project or within a phase of a larger project.

Reference: *PMBOK® Guide*, 4th Ed., page 341

174. D. The failure of the prototype test may not be the fault of the seller. At this project junction, it is important to review the contract's terms and conditions for early termination to determine the appropriate course of action.

Reference: The *PMBOK® Guide*, 4th Ed., page 341

175. D. A well-written contract normally will stipulate the terms and conditions on how to deal with unresolved claims or issues. It is important, therefore, to review the contract prior to proceeding with any further actions.

Reference: *PMBOK® Guide*, 4th Ed., pages 341–342

176. D. If at all possible, it is usually more cost effective for both parties to avoid legal actions in case of contract disagreements. Taking the issue to binding arbitration, therefore, is the least desirable outcome because it is a legal action.

Reference: *PMBOK® Guide*, 4th Ed., pages 341–342

177. A. Inputs of the Close Procurements process include the project management plan and procurement documentation.

Reference: *PMBOK® Guide*, 4th Ed., page 343

178. B. The Close Procurements process uses procurement audits, negotiated settlements, and records management systems. The other options include tools and techniques for the wrong Project Procurement Management process.

Reference: *PMBOK® Guide*, 4th Ed., page 343

179. D. Tools and techniques of the Close Procurements process include procurement audits, negotiated settlements, and records management system.

Reference: *PMBOK® Guide*, 4th Ed., page 343

180. A. The Close Procurements process produces two outputs: closed procurements and organizational process assets updates. This process often updates the procurement file, deliverable acceptance, and Lessons Learned documentation.

Reference: *PMBOK® Guide*, 4th Ed., page 344

181. A. The next process after Administer Procurements is Close Procurements. The Close Procurements process produces closed procurements and organizational process assets updates as outputs.

Reference: *PMBOK® Guide*, 4th Ed., page 344

182. D. Within the Project Procurement Management Knowledge Area processes, the Administer Procurements and Close Procurements processes update the organizational process assets. The other options are incorrect.

Reference: *PMBOK® Guide*, 4th Ed., page 314

183. C. The Code of Ethics and Professional Conduct applies to

- All PMI® members,
- Nonmembers who hold PMI® certification
- Nonmembers who have begun to apply for a PMI® certification
- Nonmembers who serve as PMI® volunteers

Reference: *Project Management Professional (PMP®) Handbook*, page 46

184. B. The Code of Ethics and Professional Conduct applies to

- All PMI® members,
- Nonmembers who hold PMI® certification
- Nonmembers who have begun to apply for a PMI® certification
- Nonmembers who serve as PMI® volunteers

Reference: *Project Management Professional (PMP®) Handbook*, page 46

185. A. The Code of Ethics and Professional Conduct applies to

- All PMI® members,
- Nonmembers who hold PMI® certification
- Nonmembers who have begun to apply for a PMI® certification
- Nonmembers who serve as PMI® volunteers

Reference: *Project Management Professional (PMP®) Handbook*, page 46.

186. D. As a project management practitioner, one of your mandatory responsibilities is to make decisions and take actions that will uphold applicable laws and regulations.

Reference: *Project Management Professional (PMP®) Handbook*, page 47

187. C. Key stakeholders should be informed in a timely manner and provided accurate information in instances where there is an experience gap or the objective has been stretched.

Reference: *Project Management Professional (PMP®) Handbook*, page 47

188. C. This is a fairly tough situational question. Politely reminding your colleagues or simply ignoring them (because no harm is done) seems appropriate; however, the Code of Ethics and Professional Conduct specifically states the following:

"We protect proprietary or confidential information that has been entrusted to us."

"We report unethical or illegal conduct to appropriate management and, if necessary, to those affected by the conduct."

Reference: *Project Management Professional (PMP®) Handbook*, page 47

189. C. The Code of Ethics and Professional Conduct states to report only ethical complaints that can be substantiated by facts.

Reference: *Project Management Professional (PMP®) Handbook*, page 47

190. D. The Code of Ethics and Professional Conduct specifically states that "We inform ourselves about the norms and customs of others and avoid engaging in behaviors they might consider disrespectful." Even though the person to whom the joke was directed was not personally offended, others might find it offensive.

Reference: *Project Management Professional (PMP®) Handbook*, page 48

191. B. The Code of Ethics and Professional Conduct specifically states that, "Fairness is our duty to make decisions and act impartially and objectively." Given that your employer may benefit from the outcomes of the motion, it is best to abstain from the discussions altogether to avoid the perception of conflict of interest.

"When we realize that we have a real or potential conflict of interest, we refrain from engaging in the decision-making process or otherwise attempting to influence outcomes, unless or until: we have made full disclosure to the affected stakeholders; we have an approved mitigation plan; and we have obtained the consent of the stakeholders to proceed."

Reference: *Project Management Professional (PMP®) Handbook*, page 48

192. D. Project management practitioners are expected to take corrective actions as appropriate if they have not been fair in their actions. Apologizing to the other party might be a correct course of action, but taking corrective actions includes that option, as well as other potential options, so it is a more correct answer.

Reference: *Project Management Professional (PMP®) Handbook*, page 48

193. A. As project managers, we are expected to be truthful in our communications and in our conduct.

Reference: *Project Management Professional (PMP®) Handbook*, page 49

194. B. One example of an aspirational standard of fairness is to demonstrate transparency when making decisions. The other options are mandatory standards of fairness.

Reference: *Project Management Professional (PMP®) Handbook*, page 48

195. D. It is a mandatory standard of honesty to not engage in dishonest behaviors. The other options are aspirational standards of honesty.

Reference: *Project Management Professional (PMP®) Handbook*, page 49

196. C. Telling half-truths or withholding some information to influence the outcome of a discussion is a violation of the mandatory standard of honesty.

Reference: *Project Management Professional (PMP®) Handbook*, page 49

197. B. All options are essentially correct except for B. Commitments must be made, implicitly and explicitly, in good faith all the time—not some of the time.

Reference: *PMBOK® Guide*, 4th Ed., page 49

198. B. Conduct that results in physical harm or creates an intense feeling of fear or humiliation is considered abusive behavior.

Reference: *Project Management Professional (PMP®) Handbook*, page 50

199. B. The best way to resolve conflicting duties of loyalty is to fully disclose the conflict. Fully disclosing the conflict can help one adhere to the duty of loyalty, thus avoiding any potential or actual conflict of interest. Avoiding the discussion altogether violates the duty of honesty.

Reference: *Project Management Professional (PMP®) Handbook*, page 50

200. A. A person's responsibility, legal or moral, to promote the best interest of an organization or other person with whom he or she is affiliated is called duty of loyalty. The other options are incorrect.

Reference: *Project Management Professional (PMP®) Handbook*, page 50

201. C. A portfolio is a collection of projects, programs, and other portfolios.

Reference: *PMBOK® Guide*, 4th Ed., page 8

202. D. In balanced and strong matrix structures, the project manager's titles are simply project managers. In a weak matrix, the title can be project coordinator, project leader, or project expeditor.

Reference: *PMBOK® Guide*, 4th Ed., page 30

203. D. Develop Project Charter is a process within the Initiating process group. Develop Project Management Plan is part of Planning, and the Direct and Manage Project Execution process is part of the Executing process group.

Reference: *PMBOK® Guide*, 4th Ed., page 43

204. A. The Verify Scope process is part of the Monitoring and Controlling process group.

Reference: *PMBOK® Guide*, 4th Ed., page 43

205. B. Interviews are usually conducted one-on-one whereas focus groups require a trained moderator.

Reference: *PMBOK® Guide*, 4th Ed., page 107

Practice Test B

1. You have been asked to initiate a business-critical project for your organization. Historically the organization has survived by creating ad hoc processes and documents, with mixed results. You want to use best practices wherever possible on this project to maximize your chances of success. Which of the following would you do first?

   A. Calculate the project timeline and initial budget.

   B. Recruit resources with the key project skills as early as possible.

   C. Develop the project charter and scope of work.

   D. Write a comprehensive change control process.

2. A project manager, new to an organization, is employed on a project to refurbish an office block. What is one of the first things that the project manager should do?

   A. Develop a project work breakdown structure.

   B. Write and get approval of a project charter.

   C. Estimate the initial budget for the project work.

   D. Produce a project schedule for the work required.

3. A project manager is developing a project charter for a project. The project is to be paid for by an external customer of the organization she works for. Which of the following items should be in place to complete the project charter?

   A. The detailed list of features to be delivered

   B. The final price quoted for the project

   C. The agreed-upon contract for the project

   D. The estimated labor costs of the project

4. As a project manager, you have been given a statement of work for a project you are assigned to manage. This project is internal to your organization. Who will have provided you with this statement of work?

   A. The end-user group

   B. Your line manager

   C. The project director

   D. The project sponsor

5. You have been given a short time to write a project charter. The project is new to your organization and will develop an IT system to help sales close leads. Which of the following would be *least* likely to help you clarify the influences on the project when writing the project charter?

   A. Escalating costs of acquiring new customers

   B. Regulatory standards that are about to be approved

   C. The infrastructure in place to support an IT solution

   D. Introduction of a new course at another university

6. You must present your project budget to the project sponsor for approval. You are anticipating some of the questions that she may ask about the project benefits to the organization. The following data shows expected cash flow for the next five years. What is the payback period for this project?

End of Year	Cash In	Cash Out
1	–	450,000
2	250,000	200,000
3	450,000	50,000
4	350,000	50,000
5	300,000	50,000

A. Two years

B. Three years

C. Four years

D. Five years

7. The project charter you have developed has not yet been approved by the project sponsor. Your direct line manager insists that you start work on the project immediately because he thinks time is tight. What is the best course of action for you?

A. Review and adapt the organization change control process.

B. Immediately begin to work on time-critical work packages.

C. State the likely impact of proceeding without approval.

D. Continue to work only on approved project work packages.

8. As part of developing a project charter, you have leveraged the company's project management methodology to help you. Which of the following categories does the project management methodology fall under?

A. International quality standards

B. Documentation management system

C. Integrated change control process

D. Organizational process assets

9. The project you have been asked to develop a charter for is complex and has many inputs that are beyond your direct knowledge. Which of the following tools and techniques would you use to help you proceed with the project chartering process?

A. Matrix management structure

B. Available project templates

C. Previous project schedules

D. Expert judgment of others

10. When planning and implementing a project, you must consider a number of influences. Which of the following factors will help you understand the project environment?

A. Schedule of all resources

B. Cultural and social issues

C. Project budget approval process

D. Detailed requirements analysis

11. You have been asked to analyze a project charter that has been developed by another part of your business. Which of the following would you consider essential for the project charter to be approved?

A. Detailed work and schedule estimates

B. A list of all the resources required

C. The business need for the project

D. A list of all the risks in the project

12. One early task the project manager should complete is to develop the work breakdown structure (WBS) for the project. This WBS will be used to do which of the following?

A. Show the constraints on the resources in the organization

B. Define and organize the total scope of the project

C. Indicate the logical dependencies between project tasks

D. Describe the scope of the new project to the customer

13. On your project, a configuration verification and audit has been scheduled. A team member asks you what this activity entails. Your response should indicate that configuration verification and audit accomplishes which of the following?

A. Involves reviewing deliverables with the customer or sponsor to ensure they are completed satisfactorily and obtaining formal acceptance by the customer or sponsor

B. Ensures that the composition of a project's configuration items is correct and that corresponding changes are registered, approved, tracked, and correctly implemented

C. Monitors and records the results of executing quality activities to assess performance and recommend necessary changes, and the configuration verification and audit is done throughout the project

D. Includes the process of tracking, reviewing, and regulating the progress to meet the performance objectives defined in the project management plan, and the configuration verification and audit is done throughout the project

14. You have to develop a project scope statement for a cross-department project. Many people will be involved in this project. Who would approve the project scope statement that you have developed?

A. The project team members

B. The project sponsors

C. The project manager

D. The financial manager

15. You have been provided with a document that contains the market demand and cost-benefit analysis that justifies the go-ahead of the project. What is this document called?

A. A contract

B. A statement of work

C. A business case

D. An organizational process asset

16. As a result of comparing planned results to actual results, change requests may be issued that expand, adjust, or reduce project or product scope. All of the following are changes that are included as part of change requests *except* ______________.

A. Corrective action

B. Situational action

C. Defect repair

D. Preventive action

17. A project manager has been given a completed project scope statement that she has not seen before. She has been asked to manage this project from now on. What is the first action she should take at this point?

A. Create a complete network diagram for the tasks and milestones in the project

B. Confirm that the project management team developed this statement

C. Develop a detailed project plan based on the new work breakdown structure

D. Call a meeting of the project management team to agree on a procurement plan

18. The project you are managing has a WBS. A new team member asks you the purpose of the WBS. What is its primary purpose?

A. To clarify the responsibility for project tasks

B. To communicate 100 percent of the project scope

C. To define the business need for the project

D. To detail the dates for the work packages

19. You are the project manager for a small project team. In one of the regular status meetings, a team member proposes an additional feature to the system you are developing. As a project manager, you remind the team to formally document the proposed additional feature. Which process is this an example of?

A. Perform Quality Control

B. Perform Change Management

C. Control Scope

D. Control Configuration

20. The organization you work for is very traditional, and projects must be managed within a functional organizational structure. Your functional manager assigned you to a project that is in trouble. What is your likely level of authority over the management of this project?

A. High

B. Moderate

C. Limited

D. Little or none

21. The project has produced a requirements traceability matrix that links requirements to their origin and traces them throughout the project life cycle. Which statement describes the purpose of this document?

A. It describes, in detail, the project's deliverables and the work required to create those deliverables and includes product and project scope description.

B. It ensures that requirements approved in the requirements documentation are delivered at the end of the project and helps manage change to product scope.

C. It is a narrative description of products or services to be delivered by the project and is received from the customer for external projects.

D. It provides the necessary information from a business standpoint to determine whether or not the project is worth the required investment.

22. You have been developing a project scope statement for a project that you have been given authority to manage. During this process, you came across areas of technical detail that you were not familiar with. Which of the following would you use to help clarify the related issues and their impact on the project scope?

A. Experienced managers

B. Special interest groups

C. Expert judgment

D. Similar project plans

23. You have been given a document for a project to review. The document contains details of how the project will be executed, monitored, controlled, and closed. What is this document called?

A. Project charter

B. Project management plan

C. Process improvement plan

D. Scope management plan

24. You have been assigned to develop a new configuration management system that should include configuration control and change control. Which of the following identifies the functions of change control?

A. Identifying, proving, approving, tracking, and validating changes

B. Submitting, approving, tracking, measuring, and validating changes

C. Identifying, requesting, assessing, validating, and communicating changes

D. Reviewing, approving, tracking, validating, and proving changes

25. You are managing the execution of a project in your organization. You work with many people and groups regularly during this stage of a project. Whom do you work most closely with to direct, manage, and execute the project?

A. The project initiator and sponsor

B. The business unit as the customer

C. No one; as project manager, you assume full responsibility

D. The project management team

26. You have been given the responsibility for a project that you have not been involved with to date. The project is at the beginning of the execution stage. What is the main document to which you would refer to guide you on what you should be doing on this project?

A. Project management plan

B. Procurement management plan

C. Communication management plan

D. Project scope management plan

27. You are asked by your line manager to describe your current activities as a project manager. You list the activities as obtaining, managing, and using resources to achieve the project objectives. In what stage is this project?

A. Initiation

B. Execution

C. Planning

D. Controlling

28. During the process of executing a project, a number of influences drive a project manager's activities. Which of the following is an input to this process?

A. Outstanding defects and faults

B. Administrative procedure edits

C. Approved change requests

D. Postponed change requests

29. You are managing a project that is time critical and essential for the survival of the business. You have a number of changes on this project. Which of the following do you spend your time scheduling into the work for the project team?

A. Likely changes to the schedule

B. Change requests to the scope

C. Requested changes to budget

D. Approved change requests

30. You are involved in a project and regularly receive regular work packages to complete from the project manager. The project manager describes some of the results of your work as project deliverables. The processes that describe project deliverables from your work are defined in which of the following?

A. The completed task contract form

B. The initial work breakdown structure

C. The Project management plan

D. The initial project scope document

31. You are managing a team for a project that is in the execution stage. A new team member has joined the team and asks you to define a project deliverable. Which of the following is the *best* answer to this question?

A. All items consumed in the project during the execution

B. The goods that are delivered to the shipping dock and are signed for

C. A product purchased according to the procurement plan

D. Something that is produced to complete the project

32. You are managing a project that is in progress. Many tasks have been completed, some are in progress, and others are yet to start. You are reviewing your workload related to the work performance monitoring of the activities in the project. When is the best time to collect this information?

A. At the start of the activity

B. Routinely and regularly

C. At the end of the task only

D. Monthly, for progress reports

33. When running a project, the project manager must manage the project work. Which of the following statements is part of effectively monitoring and controlling the project progress?

A. Email the team the schedule according to the plan.

B. Record the actual progress on task on a daily basis.

C. Compare actual activity performance against the plan.

D. Report only completed activity, schedule, and costs.

34. One of the tools you use during project management is the earned value management technique. Your line manager asks you to justify the time you spend on reporting progress. Which of the following statements *best* explains why you are using the earned value management technique?

A. Future performance can be made to exactly meet the plan.

B. Past performance can be measured to an accuracy of less than 5 percent.

C. Future performance can be predicted to within 10 percent of the budget.

D. Future performance can be forecasted based on past performance.

35. Which of the following statements *best* describes the purpose of using preventive actions in a project?

A. It reduces the probability of negative consequences related to risks.

B. It reduces the impact of positive consequences related to risks.

C. It increases the project budget to allow for some overrun on costs.

D. It increases the reporting frequency on activities that are critical.

36. You are planning a project and want to introduce the concept of integrated change control to your team. The team is used to a casual management style, and you anticipate some questions. What is the best way to describe the use of integrated change control to the team?

A. Change control applies to the inception stage to define the scope only.

B. Change control applies from the beginning of the project to the end.

C. Change control is used only when there are large changes to budget.

D. Change control is used at the execution stage to control schedule creep.

37. You are managing a section of a large project and have adopted an integrated change control system for the constant flow of changes from the project team. Which of the following describes how you would act on changes?

A. Approve changes that cost less than 10 percent of budget.

B. Automatically approve all changes to the schedule.

C. Review, then approve or reject, project changes.

D. Reject all changes to the budget, scope, or deliverables.

38. You are asked to describe why you are using the Perform Integrated Change Control process in your projects. You refer to the systems and processes that are in place in your workplace. Which of the following statements *best* describes your reasons for using change control?

A. There is a form that is always used on every project.

B. Change control is always used to limit budget overspend.

C. It is required on the entire project because of legislation.

D. It is part of the project management methodology.

39. A project that is using the Perform Integrated Change Control process will have a number of outputs. A member of your team suggests that the project management plan is an output of this process. What should your answer be?

A. Disagree because updates to the project management plan are outputs

B. Disagree because the project management plan is not an output

C. Agree that the project management plan is an output

D. Disagree because project management plan is not used in this process

40. Your organization has implemented best practices in project management processes. One function that has been set up is the Project Management Office (PMO). What is the function of the PMO?

A. To provide hot desk facilities for all project managers

B. To coordinate the management of projects

C. To close project accounts at the end of the projects

D. To provide standard stationery for project paperwork

41. You are at the end of a project that has been successfully delivered. One of the processes that you have to manage is called the Close Project or Phase process. You are not certain what this entails. Where would you get guidance on the process to help you close the project?

A. The procurement management plan

B. The project scope management plan

C. The project management plan

D. The quality management plan

42. You are managing the Close Project or Phase process for a project that you have delivered. You must gather information to allow you to complete this process. Which of the following is an input to this process?

A. Final product/service

B. Project management plan updates

C. Change request status updates

D. Accepted deliverables

43. The project management team is performing the Collect Requirements process. As part of the process, a selected number of experts are answering questionnaires and providing feedback on the responses from each round of requirements gathering. To maintain anonymity, the source of the responses is not available to the facilitator. What is this technique called?

A. Observation technique

B. Facilitated workshops

C. Delphi technique

D. Monte Carlo technique

44. The project management team has produced a requirements management plan. What is the purpose of this document?

A. To link requirements to their origin and trace them throughout the project

B. To document how requirements will be analyzed, documented, and managed throughout the project

C. To describe how individual requirements meet the business need for the project

D. To provide guidance on how project scope will be defined, documented, verified, managed, and controlled

45. You are responsible for the Close Project or Phase process on a large project. You are asked to report the activities you are performing. These activities include defining how the project will transfer the services produced to the operational division of the organization. What term describes these activities?

A. Contract closure

B. The Project management plan

C. Administrative closure

D. Deliverable acceptance

46. You have been managing a project for some time and have been asked to identify the current stage of the project. You have a document that formally indicates that the sponsor has officially accepted the project deliverables. Which of the following documents is this considered a part of?

A. Administrative closure documents

B. Contract closure documents

C. Project closure documents

D. Organizational process assets

47. What is the collection of documents that includes the project management plan, cost baseline, project calendar, and risk register called?

A. The project files

B. Historical information

C. Acceptance documents

D. Closure documentation

48. You have been given the responsibility of documenting why a project was terminated. Which of the following documents would you use to record this information?

A. Project contract document

B. Project closure document

C. Contract closure procedure

D. Administrative closure procedure

49. You are discussing your role as a project manager with your peers. In the conversation, another team member describes the process of defining and controlling what is, and what is not, included in the project. What is this process called?

A. Project documentation management

B. Project change control

C. Project scope management

D. Formal acceptance documents

50. You have been involved in a workshop at which the project management team created a work breakdown structure (WBS). Which of the following statements *best* describes the process in which you were involved?

A. Calculating the total duration of the project from the start

B. Counting the total number of work packages in the project

C. Allocating responsibilities for the project work to individuals in the team

D. Subdividing the deliverables into smaller, more manageable components

51. The scope management plan is produced by the project management team as part of the Develop Project Management Plan process. Which of the following statements *best* describes the scope management plan?

A. The scope management plan provides guidance on how project scope will be defined, documented, verified, managed, and controlled.

B. The scope management plan documents how requirements will be analyzed, documented, and managed throughout the project.

C. The scope management plan describes, in detail, the project's deliverables and the work required to create those deliverables.

D. The scope management plan is a deliverable-oriented hierarchical decomposition of the work to be executed by the project team.

52. A team member who has just joined your project asks what is involved in this project. You refer her to the project scope statement, which contains all of the following *except* ______________.

A. Deferred change requests

B. Product scope description

C. Project assumptions

D. Product acceptance criteria

53. Which of the following descriptions about requirements documentation are *false*?

A. Must be measurable, testable, and traceable

B. Serves as working model of the project's product

C. Can start at a high level and then be refined later

D. Includes a simple list of prioritized requirements

54. You are assigned to manage a new project. Your line manager suggests that you should use a tool to help you plan the new project. She suggests that you use a work breakdown structure (WBS) template. Which of the following is the *best* description of this tool?

A. A document that lists WBS elements

B. An all-purpose WBS from the Internet

C. A WBS from a previous project

D. A definition of WBS colors and shapes

55. The project meeting you are attending is becoming heated, and arguments are starting about the work completed and the work to be done. The project manager stops the discussion and refers to one document that is used as a reference. What is this document called?

A. Approved changes

B. Newly identified risks

C. The WBS template

D. The scope baseline

56. The project team you are working with is doing the work of obtaining formal stakeholders' acceptance of the completed project and associated deliverables. Each deliverable is reviewed against the requirements to check that it has been completed satisfactorily. What is this process called?

A. Verify Scope

B. Define Scope

C. Control Scope

D. Scope Management

57. You are discussing your project roles with a colleague. She states that she is preoccupied with ensuring that all scope changes go through integrated change control. What project process is she performing?

- **A.** Define Scope
- **B.** Control Scope
- **C.** Identify Risks
- **D.** Plan Procurements

58. The project team is involved in measuring, examining, and verifying that the work and deliverables meet the product and acceptance criteria. Which of the following is a summary description of the work the team is doing?

- **A.** Define Scope: Inputs
- **B.** Verify Scope: Outputs
- **C.** Verify Scope: Inspection
- **D.** Scope Management: Inputs

59. The Control Scope process is part of the ______________ knowledge area under the ______________ process group.

- **A.** Project Integration Management; Monitoring and Controlling
- **B.** Project Quality Management; Planning
- **C.** Project Risk Management; Executing
- **D.** Project Scope Management; Monitoring and Controlling

60. A configuration management system with an integrated change control includes two main components: ______________, which deals with both deliverables and processes, and ______________, which ensures proper administration of the changes in project and product baselines.

- **A.** Configuration control; change control
- **B.** Change requests; change control
- **C.** Configuration control; change requests
- **D.** Change control; configuration control

61. A number of changes to the project schedule and requirements have been suggested by the project team and the customer. These are urgent, and all requested changes impact the resources you are using. What is the recommended next step to take?

- **A.** Have the changes go through the Perform Integrated Change Control process for review.
- **B.** Notify the project sponsor that the work will start immediately.
- **C.** Ignore the changes to schedule and requirements.
- **D.** Reschedule resources to begin the changes immediately.

62. After some changes to your project scope, you have revised the schedule, costs, and work package assignments, and these changes have been approved. What action do you now complete?

A. Wait for resources to complete current assignments

B. Update the project management plan

C. Immediately look for other changes to submit

D. Let the project schedule continue as previously

63. As you plan a project assigned to your team, you identify the activities, dependencies, and resources needed to produce the project deliverables. What is this knowledge area called?

A. Project Schedule Control

B. Project Risk Management

C. Project Time Management

D. Project Cost Planning

64. You have asked a team member to participate in performing a number of actions described within the project management plan. These actions involve decomposing the work packages into smaller components of work called activities. What process is this team member responsible for?

A. Schedule Work

B. Estimate Tasks

C. Schedule Activities

D. Define Activities

65. You note that your weekly time sheet report shows that you have identified activities, scoped the work needed on each, and provided sufficient detail to allow a team member to understand what is required to complete that work. What is a summary of the time you have spent this week?

A. Producing an activity list

B. Defining the project scope

C. Developing the project schedule

D. Identifying the WBS elements

66. You are analyzing a project schedule on which activities are represented on a diagram as boxes and the dependencies between activities are shown as arrows. What is this representation known as?

A. Activity on arrow

B. Activity on node

C. Schedule analysis

D. Critical path method

67. The precedence diagramming method can be used to show the dependencies between activities. One type of dependency used is to ensure that the successor activity does not start until the completion of the predecessor activity. What is the name given to this type of dependency?

A. Start-to-start

B. Finish-to-finish

C. Finish-to-start

D. Start-to-finish

68. Which of the following is a project management deliverable that identifies and describes the types and quantities of each resource required to complete all project work packages?

A. Resource calendar updates

B. Activity attribute updates

C. Resource breakdown structure

D. Activity resource requirements

69. You are completing the planning for a project schedule. Because you do not have much information about one particular activity, you decide to estimate its duration by referring to the actual duration of a similar activity on another project. What is this calculation method called?

A. Analogous estimating

B. Expert judgment

C. Parametric estimating

D. Reserve analysis

70. Your project manager asks you to calculate the theoretical early start and finish dates, along with late start and finish dates, for all the project activities. He suggests you use a forward and backward pass analysis. What is this technique known as?

A. Critical chain method

B. Critical path method

C. Schedule compression

D. What-if scenario analysis

71. You are asked to review a project that another department has planned for your organization. Your line manager asks you to show her a project overview that displays only the start and end dates of major deliverables, along with key external dependencies. This graphic is known as a ______________.

A. Network diagram

B. Summary bar chart

C. Milestone chart

D. Schedule baseline

72. You are reviewing the project schedule with your team. One of the schedules lists the activity identifier, activity description, and calendar units. The start and end date of each activity is depicted by a horizontal bar, which clearly shows the activity duration relative to the project schedule time frame. There is also a vertical line marked "Data Date." What is this graphical representation of the schedule called?

A. Milestone schedule

B. Network diagram

C. Detailed schedule with logical relationships

D. Summary schedule

73. Which of the following statements is correct about a schedule baseline?

A. Developed from the schedule network analysis, the project management team approves the schedule baseline based on specific baseline start and end dates.

B. The schedule baseline is the basis from which the project management team calculates the project completion date based on optimistic, most likely, and pessimistic estimates.

C. Reviewed and approved by the project sponsor, the schedule baseline includes the corporate calendar, project calendar, resource calendar, and a list of statutory holidays.

D. The project calendar is exactly the same as a regular calendar except that Saturdays are always used in the preparation of the overall project schedule.

74. The project team used schedule network analysis, the critical path method, the critical chain method, and other tools to produce the project schedule and schedule baseline. These tools and techniques, and the resulting outputs, compose the ______________ process.

A. Define Activities

B. Develop Schedule

C. Control Schedule

D. Estimate Activity Durations

75. As a project manager, your responsibilities include gathering data on actual start and finish dates of activities, along with remaining durations for work in progress. The work you are doing is involved with ______________.

A. Cost variance analysis

B. Performance measurement

C. Performance review

D. Critical path analysis

76. What does the term *variance analysis* mean for a project manager, in reference to the schedule?

A. Recording the actual start date of critical activities in the project

B. Analyzing the calendar contract start date and finish date

C. Calculating the difference between total slack and free slack

D. Comparing activity target start date with actual start date

77. You are managing a project that is running late. You have proposed corrective actions that will affect the schedule baseline. What are these actions called?

A. Change requests

B. Action change control

C. Schedule baseline updates

D. Project scope updates

78. The project archive you are reviewing shows that the project has had a number of changes to the start and finish dates of the schedule baseline. What are these changes known as?

A. Schedule variance analysis

B. Schedule baseline updates

C. Approved change requests

D. Performance reviews

79. Which knowledge area includes processes that ensure the completion of the project within the approved budget?

A. Project Risk Management

B. Project Schedule Management

C. Project Cost Management

D. Project Resource Management

80. You have been working on a project in the planning stage, developing the cost estimates. The project sponsor expects a narrower estimate of the costs because more information is now known. Which of the following is considered a suitable range for this request?

A. –100 percent to +100 percent

B. –10 percent to +10 percent

C. –25 percent to +25 percent

D. –50 percent to +50 percent

81. A project team member is building the costs for a current project, based only on costs from a previous similar project that has been completed in your organization. What is this process known as?

A. Analogous estimating

B. Bottom-up estimating

C. Parametric estimating

D. Three-point estimating

82. Building cost estimates for a current project using metrics from the current project, along with historical information from past similar projects, is known as ______________.

A. Analogous estimating

B. Bottom-up estimating

C. Parametric estimating

D. Three-point estimating

83. What factors are needed to calculate the estimate to complete (ETC) for a project?

A. EAC and PV

B. BAC, AC, and PV

C. BAC, EV, and AC

D. EAC and AC

84. You have been asked to report to the project sponsor on project performance. Which of the following would help you provide this information?

A. Cost performance index (CPI)

B. Schedule performance index (SPI)

C. Estimate to complete (ETC)

D. Estimate at completion (EAC)

85. Which value is used as part of the CV, SV, CPI, and SPI calculations?

A. Planned value

B. Actual cost

C. Earned value

D. Schedule variance

86. A new team member is monitoring and recording results of executing quality activities to assess performance and recommend necessary changes. He asks for clarification on what process his activities are related to. Which of the following is the *best* answer to his question?

A. Perform Quality Control

B. Perform Quality Assurance

C. Plan Quality

D. Quality Improvement

87. A team member has been given the task of identifying which quality standards are relevant to this project and determining how to satisfy these standards. In which of the following activities is she engaged?

A. Perform Quality Control

B. Perform Quality Assurance

C. Plan Quality

D. Quality Improvement

88. As a project manager you must audit the quality requirements and the results from quality control measurements to ensure that appropriate quality standards and operational definitions are used. This is part of which of the following processes?

A. Perform Quality Control

B. Perform Quality Assurance

C. Plan Quality

D. Quality Improvement

89. You have been asked to detail the steps for analyzing processes to identify activities that enhance the value of those processes. This is called ______________.

A. Perform Quality Assurance

B. Develop Quality Checklists

C. Establish Quality Metrics

D. Perform Process Improvement

90. You have been asked to perform quality planning for a project. One of your first actions will be to clarify some of the project or product attribute measurements on the project and how the Perform Quality Control process will measure these attributes. The output of the Plan Quality process as described here is known as ______________.

A. Quality metrics

B. Quality checklists

C. Quality management plan

D. Process improvement plan

91. Which of the following is an input to Perform Quality Assurance?

A. Quality metrics

B. Project requirements

C. Change requests

D. Validated deliverables

92. A project manager is reviewing a project that is in progress. She is trying to identify good project practices being used and determine gaps between best practices and current practices. In addition, she is sharing the good practices with similar projects in the organization. This tool or technique is called _____________.

A. Quality metrics

B. Quality audit

C. Quality improvement

D. Quality checklists

93. A team member is following the steps outlined in the process improvement plan. This is needed to improve organizational and project performance. What technique will the team member use to help with this activity?

A. Quality metrics

B. Quality audits

C. Process analysis

D. Quality checklists

94. In the Perform Quality Control process, when is time normally allocated to do this work?

A. At every project milestone

B. At termination of the project

C. At the initiation stage only

D. Throughout the project

95. You are coaching a new team member in your project. One of her functions is to recommend corrective actions to increase the effectiveness and efficiency of the organization. This falls under the Perform Quality Assurance process. This list of actions that results from carrying out this process is documented as _____________.

A. Change requests

B. Organizational process assets updates

C. Project management plan updates

D. Recommended preventive actions

96. The project you are managing has a serious problem that will compromise the delivery of a key component. Your team is trying to find the reasons for the failure of the system, using a tool consisting of a diagram that shows how various factors might be linked to the problem or the effects. What is this diagram called?

A. Variable scatter diagram

B. Cause and effect diagram

C. Process control chart

D. Statistical sampling

97. Your mentor suggests that you use a chart that shows, by frequency of occurrence, events and defects in the project. What is the common name for this chart?

A. Scatter diagram

B. Flowchart

C. Pareto chart

D. Control chart

98. A common tool used in analyzing problems on a project is a diagram that shows the relationship between two variables. What is this diagram called?

A. Control chart

B. Pareto chart

C. Run chart

D. Scatter diagram

99. You are identifying and documenting project roles, responsibilities, and reporting relationships for a project, as well as creating a staffing management plan for a project. Which process is responsible for these activities?

A. Develop Human Resource Plan

B. Develop Project Management Plan

C. Manage Project Team

D. Develop Project Team

100. As part of the project plan documentation, you find a chart that illustrates the link between work packages and project team members. What is this chart called?

A. Hierarchy-type chart

B. Responsibility assignment matrix

C. Organizational chart

D. Network diagram

101. You are planning the staff for a project. One of the diagrams you produce illustrates the number of hours that each person will be needed each week over the course of the project schedule. What is this chart commonly known as?

A. Work breakdown structure

B. Task network diagram

C. Resource histogram

D. Detailed Gantt chart

102. Your team has proposed that a resource-leveling strategy should be adopted for the project schedule from this point on. Which of the following statements *best* describes what they are suggesting?

A. No change to the resources and reducing the project schedule

B. Adding resources and reducing the project schedule

C. Reducing resources but not changing the project schedule

D. Balancing out the peaks and valleys of resource usage

103. You meet with the project sponsor, and the conversation covers the following topics: availability, competencies, experience, interests, and costs of the potential project team. The items referenced are part of which of the following?

A. Enterprise environmental factors

B. Project resource histogram

C. Responsibility assignment matrix

D. Organizational breakdown structure

104. Some of the project staff assignments are defined within the project charter. What is this assignment of tasks known as?

A. Resource management

B. Preassignment

C. Resource leveling

D. Responsibility assignment matrix

105. You are managing a team whose members all have a shared goal, and they all perform their roles, spending little or no time meeting face to face. What is this type of team environment called?

A. Composite organization

B. Matrix organization

C. Virtual team

D. Remote team

106. You are managing a project and ask for additional time to hire outside resources. You will need to prepare the job descriptions before you can hire individual contractors or subcontract the work to a third party. This scenario typically describes a tool or technique of which process?

A. Conduct Procurements

B. Close Procurements

C. Develop Project Team

D. Acquire Project Team

107. You have been asked to give details of how you propose to develop the project team. Which of the following descriptions *best* describes what you should be doing?

A. Improve the competencies of team members

B. Document the resource calendars of team members

C. Assign the appropriate people to activities on the project

D. Finalize roles and responsibilities in the project plan

108. Which process uses the following tools and techniques: interpersonal skills, team-building activities, and co-location?

A. Identify Stakeholders

B. Develop Project Team

C. Develop Human Resource Plan

D. Manage Stakeholder Expectations

109. At your personal performance review, your line manager suggests a number of areas in which you can improve. These include understanding better the sentiments of the project team members and acknowledging their concerns. These skills are *most* related to _______________.

A. Team building

B. Ground rules

C. Interpersonal skills

D. Co-location

110. Which of the following tools and techniques is *not* part of developing a project team?

A. Interpersonal skills

B. Training

C. Team-building activities

D. Conflict management

111. You are preparing a budget item for your project to support a facilitated offsite meeting designed to improve interpersonal relationships within the project team. The title for this budget line item should be ______________.

A. Team-Building Activity

B. Soft Skills Training

C. Effective Team Working

D. Improving Team Competencies

112. You are required to develop a recognition and reward system for your project office. Which of the following is an appropriate basis for this system?

A. Reward only desirable behavior

B. Reward the team member of the month

C. Reward all team members who work overtime

D. Reward individualism regardless of culture

113. Which of the following organizational structures creates a more complicated environment in managing the project team?

A. Functional organization

B. Matrix organization

C. Projectized organization

D. Hierarchical organization

114. Tracking team performance, providing feedback, resolving issues, and coordinating changes to enhance project performance are all part of which process?

A. Develop Project Team

B. Acquire Project Team

C. Manage Project Team

D. Develop Human Resource Plan

115. Measuring the effectiveness of the project team, recording current competencies, and monitoring reduction in team turnover rates are related to ______________.

A. Project staff assignments

B. Project team-building activities

C. Recognition and reward systems

D. Team performance assessment

116. Schedule control, cost control, quality control, scope verification, and procurement audits are all examples of inputs to managing the project team known as ______________.

A. Project performance reports

B. Team performance assessment

C. Staffing management plan

D. Organizational process assets

117. Certificates of appreciation, bonus structures, and corporate apparel are ______________ that can influence the management of the project team.

A. Staff roles and responsibilities

B. Organizational process assets

C. Staffing management plan

D. Team performance assessment

118. To help you manage the project team, you are documenting who is responsible for resolving a specific problem by a target date. The project team will *most likely* refer to the resulting document as a ______________.

A. Project risk register

B. Change control

C. Project issue log

D. Performance report

119. A project manager must perform a number of activities when planning and managing a project. These include identifying and determining the information needs of project stakeholders, making information available in a timely manner, collecting and distributing performance information, and meeting information needs to help resolve issues with stakeholders. These activities are considered part of which of the following knowledge areas?

A. Project Scope Management

B. Project Human Resource Management

C. Project Integration Management

D. Project Communications Management

120. Which of the following are inputs and outputs of the Plan Communications process?

A. Stakeholder register and stakeholder management strategy, communications management plan, and project document updates

B. Stakeholder register and stakeholder management strategy, enterprise environmental factors, and organizational process assets

C. Communication requirements analysis and communication technology, communication models, and communication methods

D. Communication models and communication methods, Communications Management plan, and project document updates

121. A section of your project plan lists methods and technologies to be used to convey memoranda, email, and press releases, and how frequently these should be used. What is the document that defines these called?

A. Communications management plan

B. Organizational assets

C. Communications technology

D. Project management plan

122. During the planning stage of a project, you assign a member of your project team to be responsible for the distribution of information about the project. This responsibility is defined in a document called (a) _______________.

A. Project management plan

B. Communications management plan

C. Project roles and responsibilities

D. Staffing management plan

123. When sending information to project stakeholders, the project manager is responsible for making the information clear and complete so that the recipient can understand it correctly. What attribute of a project manager does this statement describe?

A. Influence skills

B. Negotiating skills

C. Communications skills

D. Delegation skills

124. When managing project stakeholders, you may use a number of different techniques, including written and oral, both giving and receiving. These techniques are examples of _______________.

A. Influence skills

B. Negotiating skills

C. Delegation skills

D. Communications skills

125. The document that covers the topics of identifying project successes and failures, and making recommendations on how to improve future performance on other projects, is referred to as the _______________.

A. Lessons Learned documentation

B. Project management plan

C. Issue log

D. Change log

126. You have completed an end-of-phase review meeting. Several actions and suggestions have been given to you as the project manager. You must document these so that they are considered for use on future projects. These actions and suggestions are called (the) ________________.

A. Project Issues list

B. Lessons Learned

C. Risk register items

D. Project documentation

127. Which of the following is the recommended way of reporting on project progress to stakeholders?

A. Regularly, at preset times

B. Comprehensively, at the end of the project

C. Regularly and on exception (ad hoc)

D. On exception (ad hoc) and at the end of project

128. You have been asked to collect and provide performance information to the project stakeholders. The process responsible for compiling reports is called ________________.

A. Baselining the schedule

B. Stakeholder management

C. Team performance assessments

D. Report Performance

129. Which of the following is an input to the Report Performance process?

A. Budget forecasts

B. Performance reports

C. Change requests

D. Organizational process assets updates

130. Work performance information that indicates the status of deliverables and what has been accomplished to date in the project is part of which process group?

A. Project Initiation

B. Project Execution

C. Project Closure

D. Project Planning

131. You have a number of documents and data, including bar charts, S-curves, histograms, and tables for the data analyzed against the performance measurement baseline. The collective name for this information is ______________.

A. Performance measurements

B. Forecast completion

C. Performance reports

D. Deliverable status

132. As part of your role as project manager, you must frequently update and reissue work performance information that has been analyzed as the project proceeds. This information concerns how the project's past performance could impact the project in the future. This information is documented as:

A. Performance reports

B. Corrective actions

C. Change requests

D. Variance analysis

133. The activity of managing communications to satisfy the needs of, and resolve issues with, all the project stakeholders is usually the responsibility of which role in a project team?

A. Project manager

B. Everyone on the team

C. Project sponsor

D. Communication specialist

134. Which of the following is a tool that a project manager uses to help document and monitor the resolution of issues in the project?

A. Risk register

B. Issues log

C. Change requests

D. Corrective actions

135. The set of project management activities including identification, analysis, planning responses, and monitoring and controlling of risks in the project is part of carrying out ______________.

A. Risk identification

B. Project risk analysis

C. Project risk management

D. Project risk mitigation

136. Most of the project risk management processes are performed as part of which process group?

A. Closing

B. Executing

C. Initiating

D. Planning

137. The project director advises you that a project should have a balance between risk taking and risk avoidance. This policy is implemented in a project using which of the following?

A. Risk responses

B. Risk analysis

C. Risk identification

D. Risk classification

138. One of the tools used to manage risk in a project is a table showing the various thresholds for different levels of risks in the project. What is this table called?

A. Risk register

B. Probability and impact matrix

C. Issues log

D. Impact scales

139. Who is responsible for identifying risks in a new project?

A. Project manager

B. Project sponsor

C. Any project personnel

D. The main stakeholders

140. What is the document that contains a list of identified risks, responses to the risks, the root cause of the risk, the risk category, and additional information about the identified risks called?

A. Risk management plan

B. Project issues log

C. Risk-related contract decisions

D. Risk register

141. As part of your responsibility for managing risks in your project, you rate risks as low, medium, or high. What tool would you typically use to define these ratings?

A. Probability impact matrix

B. Risk register updates

C. Assumption analysis

D. Checklist analysis

142. A new member of your project team suggests that you use a specific technique for quantitative risk analysis. She says that you should calculate information on the lowest, highest, and most likely costs of the WBS elements in the project plan. This technique is an example of _______________.

A. Three-point estimates for costs

B. Probability and impact assessment

C. Probability distributions

D. Sensitivity analysis

143. What is the process of project planning that involves developing options, determining actions to enhance opportunities, and reducing threats to project objectives called?

A. Perform qualitative risk analysis

B. Plan risk responses

C. Perform quantitative risk analysis

D. Probability impact matrix

144. You have been given responsibility for developing a risk register for your project. Through which process is this document created?

A. Plan Risk Responses

B. Perform Quantitative Risk Analysis

C. Identify Risks

D. Perform Qualitative Risk Analysis

145. A common risk management strategy is to shift the negative impact of a threat, and ownership of the response, to a third party. What is this technique called?

A. Exploit

B. Avoid

C. Mitigate

D. Transfer

146. Your team is developing a part of the risk management plan. For some of the risks, the team decides that a response plan will be executed only when certain predefined conditions exist. What is the term given to this type of risk strategy?

A. Contingent

B. Sharing

C. Exploit

D. Enhance

147. You are managing risk of a project using the risk management plan. How often should you monitor the project work for new and changing risks?

A. At the beginning of project planning

B. Continuously throughout the project lifecycle

C. At the beginning of each project phase

D. At the end of each project phase

148. Information such as identified risks, risk owners, agreed responses, actions, warning signs, and a watch list of low-priority risks are examples from which of the following?

A. Risk management plan

B. Approved change requests

C. Risk register contents

D. Work performance information

149. Identifying and documenting the effectiveness of risk responses in dealing with identified risks and the root causes, and of the effectiveness of the risk management process, is called (a) ______________.

A. Risk mitigation

B. Risk identification

C. Risk analysis

D. Risk audit

150. A consultant has been reviewing outputs that resulted from carrying out the Monitor and Control Risks process. She lists a number of actions that are required to bring the project into compliance with the project management plan. What are these actions called?

A. Recommended preventive actions

B. Risk register updates

C. Recommended corrective actions

D. Project management plan updates

151. You are part of a team running a complex project spanning a number of years and involving a number of subcontractors. The contracts signed for this project might or will end ______________.

A. During any phase of the project

B. During the completion phase

C. During the execution phase

D. During the acceptance phase

152. During the Plan Procurements process for a project, you make a decision based on whether a particular product or service can be produced by the project team or instead should be purchased. This decision-making process is called ______________.

A. Procurement management

B. Expert judgment

C. Risk management

D. Make-or-buy analysis

153. The contract you are negotiating with a subcontractor involves a set price for a well-defined requirement. There are also incentives for meeting selected objectives. This term is a ______________.

A. Time and material contract

B. Cost-reimbursable contract

C. Fixed-price contract

D. Cost plus incentive fee contract

154. The procurement-related processes involve asking such questions as how well the seller meets the contract statement of work, whether the seller has the capacity to meet future requirements, and whether the seller can provide references from previous customers. What is the name given to this list and/or its use?

A. Contract management

B. Source selection criteria

C. Make-or-buy decisions

D. Contract negotiation

155. What is the procurement process that involves obtaining seller responses, selecting a seller, and awarding a contract called?

A. Qualified sellers list

B. Bidder conference

C. Conduct Procurements

D. Select acceptable seller

156. A list of prospective and previously qualified sellers and information on relevant past experience with sellers, both good and bad, generally is known as ______________.

A. Procurement management plan

B. Procurement documents

C. Organizational process assets

D. Qualified sellers list

157. The project sponsor asks you to ensure that all prospective sellers have a clear and common understanding of the procurement you require. He suggests you have a structured meeting to do this. What is this meeting called?

A. Change control meeting

B. One-on-one meeting

C. Project status meeting

D. Bidder conference

158. The document containing information on specific prescreened sellers, and including results of visited reference sites and reference contacts for some of the sellers' previous customers, is called (a) ______________.

A. Qualified sellers list

B. Procurement document

C. Organizational process asset

D. Procurement statement of work

159. The sellers in a project have prepared documents that describe their ability and willingness to provide the requested products, services, or results detailed in the procurement documents. What is the generic name for the documents that the sellers provided?

A. Requirements

B. Responses

C. Requests

D. Proposals

160. Your contract and procurement consultant recommends that you establish source selection criteria for each of your sellers' proposals. What part of the procurement process is this a part of?

A. Bidder conference

B. Plan Procurements

C. Conduct Procurements

D. Administer Procurements

161. Within the Conduct Procurements process, the project management plan, seller proposals, and qualified seller list are examples of ______________.

A. Contract types

B. Conduct Procurements tools

C. Conduct Procurements inputs

D. Conduct Procurements outputs

162. In a project where there are a number of sellers, the responses to a procurement document are called ______________.

A. Proposals

B. Presentations

C. Requests

D. Bidding

163. You have commissioned a consulting organization to check on the proposed pricing for a new contract. You request that the report highlight the significant differences between the pricing given by the consultants and the internal estimate. This technique makes use of ______________.

A. Independent estimates

B. Bidder conference

C. Screening system

D. Expert judgment

164. You have been assigned to a research and development project that will commercialize a new technology. Because of the high risks associated with new technologies, your company has partnered with another company to execute the project. The two companies have signed a legal contractual document that predetermines the buyer and seller roles, scope of work, competition requirements, and other issues. This document is called a ______________.

A. Cost-reimbursable contract

B. Teaming agreement

C. Noncompete agreement

D. Boilerplate agreement

165. A mutually binding legal agreement that obligates the seller to provide the specified services and also obligates the buyer to pay the seller is called a ______________.

A. Request

B. Tender

C. Procurement contract

D. Proposal

166. The specific technique for evaluating sellers' proposals that involves evaluation of the proposals by a multidiscipline team including members with specialized knowledge in each of the areas covered by the proposed contract is called (a) ____________.

- **A.** Seller rating system
- **B.** Expert judgment
- **C.** Screening system
- **D.** Bidder conference

167. The procurement management plan for a project is a subset of (the) ____________.

- **A.** Procurement documentation
- **B.** Project management plan
- **C.** Claims administration
- **D.** Qualified sellers list

168. You have been asked by the project sponsor to ensure that the seller's performance meets the contractual requirements and that your organization, as the buyer, performs according to the contract. The process responsible for these activities is called ____________.

- **A.** Administer Procurements
- **B.** Plan Procurements
- **C.** Conduct Procurements
- **D.** Selected sellers

169. The procurement management plan for a project describes how the procurement processes will be managed ____________.

- **A.** At the Seller Selection stage
- **B.** Throughout the project lifecycle
- **C.** Throughout the lifetime of the contract
- **D.** During the contract documentation stage

170. The project you are managing has a problem that requires the contract with a seller to be modified. The alteration to the contract is in accordance with the change control terms of the contract and project. The recommended time to make this change to the contract is ____________.

- **A.** Never; the contract cannot be modified at any time
- **B.** At any time regardless of the response from the seller
- **C.** At any time prior to the contract being awarded to the seller
- **D.** At any time prior to contract closure by mutual consent

171. You are managing a system that includes the following information: contract documentation, tracking systems, dispute resolution procedures, and approved levels of authority for changes. What is this system called?

A. Source selection criteria

B. Integrated change control system

C. Procurement management plan

D. Contract change control system

172. As part of the Administer Procurements process, contested changes will arise where the buyer and seller cannot reach an agreement on compensation for the change, or cannot agree that a change has occurred. These are called claims, disputes, or appeals. If the parties cannot resolve a claim by themselves, it may need to be resolved using what method?

A. Alternative dispute resolution

B. Compromising technique

C. Integrated change control

D. Economic price adjustment

173. A document produced by the buyer that rates how well each seller is performing the project work is called (a) ______________.

A. Source selection criteria

B. Seller performance evaluation

C. Procurement management plan

D. Work performance information

174. The project you have been working on is near completion. One process that you must manage as the project manager involves verification that all the work and deliverables supplied by the seller were acceptable. This process is called ______________.

A. Close Project

B. Close Procurements

C. Lessons Learned

D. Performance reporting

175. In a project, the seller is performing below the contracted level of work consistently. What is the most appropriate procedure to follow, based on the options provided?

A. Continue until the seller provides an explanation

B. Add time to the project schedule

C. Terminate the seller's contract early

D. Increase the budget allocated to the contract

176. When managing a contract in a project, the process that involves archiving the seller performance for future use is called ______________.

A. Close Project or Phase

B. Performance reporting

C. Lessons Learned

D. Close Procurements

177. The project office wants to do a structured review on your project of the procurement-related activities, from the Plan Procurements, Conduct Procurements, and Administer Procurements processes. This review is called (a) ______________.

A. Performance reporting

B. Contract management

C. Procurement audit

D. Claims administration

178. You are about to perform the Close Procurements process on a project. Which of the following is a key input to that process?

A. Procurement documentation

B. Closed procurements

C. Deliverable acceptance

D. Lessons Learned

179. What is the objective of the use of the procurement audit process on a project when conducted at the contract closure stage?

A. Identify when legal action should be started

B. Terminate the nonperforming suppliers' contracts

C. Identify who signed the nonperforming contracts

D. Identify success and failure for use in future contracts

180. Your project is near completion. You have been authorized to provide the seller of one of the project deliverables with formal notice that the contract has been completed. What process is this?

A. Lessons Learned

B. Close Procurements

C. Administer Procurements

D. Deliverable acceptance

181. Your project is near completion. You have been authorized to provide the seller of one of the project deliverables with formal notice that the deliverables have been accepted. The action that you took is called _______________.

A. Lessons Learned

B. Procurement management

C. Deliverable acceptance

D. Contract management

182. As part of the Close Procurements process, you have provided a list of recommendations for process improvements for future contracts and purchasing. What is this called?

A. Contract documentation

B. Procurement audit

C. Lessons Learned

D. Claims administration

183. In one of your project team meetings, you mentioned that practitioners in the project management profession use the *Code of Ethics and Professional Conduct* as a reference point for acceptable professional behavior. In this context, the statement that all project managers are committed to doing what is right and honorable refers to _______________.

A. Persons to whom the code applies

B. Vision and purpose

C. Structure of the code

D. Values that support the code

184. In the project management plan, there is a reference to the *Code of Ethics and Professional Conduct*. What is the purpose of the *Code of Ethics and Professional Conduct* in project management?

A. Keep unqualified project managers out of contracts

B. Enable prosecution of poorly performing project managers

C. Identify project managers who are not qualified

D. Instill confidence in the project management profession

185. What was the basis used by practitioners in the project management community to develop the *Code of Ethics and Professional Conduct*?

A. Experience from previous budget negotiations

B. Contract law as practiced in participating countries

C. International legal standards and procedures

D. Values that formed decision making and guided actions

186. With reference to the *Code of Ethics and Professional Conduct*, each section has a set of standards. These standards of conduct are described by which of the following pairs of terms?

A. Aspirational and mandatory

B. Aspirational and legal

C. Mandatory and contractual

D. Aspirational and contractual

187. As a project manager, you are expected to take ownership for the decisions you make or fail to make, the actions you take or fail to take, and the resulting consequences. What best fits this description of the conduct of a project manager, as defined in the Project Management Institute's *Code of Ethics and Professional Conduct*?

A. Responsibility

B. Respectability

C. Confidentiality

D. Consistency

188. Project management practitioners may bring violations of the *Code of Ethics and Professional Conduct* to the attention of the appropriate body for resolution. Doing so refers to what standard?

A. Mandatory responsibility

B. Aspirational responsibility

C. Aspirational respectability

D. Optional responsibility

189. It is the duty of project managers to show high regard for themselves, for others, and for the resources entrusted to them. The Project Management Institute's *Code of Ethics and Professional Conduct* refers to this behavior as ______________.

A. Honesty

B. Responsibility

C. Respect

D. Fairness

190. In project meetings, you often take time to listen to others' points of view, seeking to understand them. Referring to the Project Management Institute's *Code of Ethics and Professional Conduct*, how would you classify this behavior?

A. Aspirational responsibility

B. Mandatory responsibility

C. Mandatory respect

D. Aspirational respect

191. You are in a project meeting, and another project manager acts in a way that is abusive to another member of the team. You decide to remind the team member, outside the meeting, that this is a breach of which professional standard?

A. Mandatory responsibility

B. Mandatory respect

C. Mandatory fairness

D. Mandatory honesty

192. It is the duty of a project manager to act impartially and objectively when making decisions and otherwise behaving. The Project Management Institute's *Code of Ethics and Professional Conduct* refers to this as the value of ______________.

A. Fairness

B. Honesty

C. Respect

D. Responsibility

193. You advise your team to behave in a manner that provides equal access to information to those who are authorized to have the information. The Project Management Institute's *Code of Ethics and Professional Conduct* refers to this behavior as following the value of ______________.

A. Aspirational respect

B. Aspirational fairness

C. Aspirational honesty

D. Aspirational responsibility

194. While negotiating a contract with a supplier, you realize that you already know one of the other parties in a professional capacity. You decide you must disclose this as a potential conflict of interest. What professional standard are you applying in this situation?

A. Mandatory honesty

B. Mandatory respect

C. Mandatory fairness

D. Mandatory responsibility

195. It is the duty of a project manager to try to understand the truth and act in a truthful manner in both communications and conduct. The Project Management Institute's *Code of Ethics and Professional Conduct* refers to such behavior as following the value of ______________.

A. Fairness

B. Responsibility

C. Respect

D. Honesty

196. When describing the culture that you try to encourage in the project team, you identify trying to create an environment in which others feel safe to tell the truth as an example of appropriate behavior. The Project Management Institute's *Code of Ethics and Professional Conduct* refers to this behavior as following the value of ______________.

A. Aspirational honesty

B. Aspirational respect

C. Mandatory honesty

D. Mandatory respect

197. While managing the contracts for a project, you become aware that a team member is engaging in dishonest behavior for personal gain. You decide that you must disclose this behavior. What professional standard are you applying in this situation?

A. Aspirational fairness

B. Mandatory honesty

C. Mandatory respect

D. Aspirational responsibility

198. Conduct by a team member that results in others feeling humiliated is an example of ______________.

A. Abusive manner

B. Allowable behavior

C. Aspirational standards

D. Mandatory standards

199. Violation of the Project Management Institute's *Code of Ethics and Professional Conduct* may result in which of the following?

A. Sanctions by the Project Management Institute

B. Legal proceedings

C. Salary deductions

D. Name and shame

200. As you introduce a new team member to the organization, you describe one of the requirements of the role of project manager as the responsibility, both legally and morally, to promote the best interests of the organization. The *Code of Ethics and Professional Conduct* describes this as (the) ______________.

A. Aspirational respect

B. Duty of loyalty

C. Aspirational standard

D. Good practice

201. What is the most common logical relationship in a precedence diagramming method (PDM)?

A. Finish-to-start (FS)

B. Start-to-finish (SF)

C. Finish-to-finish (FF)

D. Start-to-start (SS)

202. The arrow diagramming method (ADM) is visually the opposite of ________________.

A. SPI

B. PDM

C. WBS

D. AOA

203. A project manager is working on a very large project for a major financial institution. While estimating the resources, he is uncertain whether to use hours, days, weeks, fixed costs, or prorated amounts. The project manager is *most likely* trying to figure out the ________________.

A. Level of accuracy

B. Unit of measure

C. Control account

D. Scope complexity

204. Which of the following is *not* an input to the Plan Communications process?

A. Stakeholder register

B. Enterprise environmental factors

C. Organization process assets

D. Communication models

205. Which risk response strategy can be used for both negative and positive risks?

A. Avoid

B. Transfer

C. Mitigate

D. Accept

Chapter 6

Answers to Practice Test B

Answer Key for Practice Test B

1. C.
2. B.
3. C.
4. D.
5. D.
6. B.
7. C.
8. D.
9. D.
10. B.
11. C.
12. B.
13. B.
14. B.
15. C.
16. B.
17. B.
18. B.
19. C.
20. D.
21. B.
22. C.
23. B.
24. C.
25. D.
26. A.
27. B.
28. C.
29. D.
30. C.
31. D.
32. B.
33. C.
34. D.
35. A.
36. B.
37. C.
38. D.
39. A.
40. B.
41. C.
42. D.
43. C.
44. B.
45. C.
46. C.
47. A.
48. B.
49. C.
50. D.
51. A.
52. A.
53. B.
54. C.
55. D.
56. A.
57. B.
58. C.
59. D.
60. A.
61. A.
62. B.
63. C.
64. D.
65. A.
66. B.
67. C.
68. D.
69. A.
70. B.
71. C.
72. D.
73. A.
74. B.
75. C.
76. D.
77. A.
78. B.
79. C.
80. B.
81. A.
82. C.
83. C.
84. B.
85. C.
86. A.
87. C.
88. B.
89. D.
90. A.
91. A.
92. B.
93. C.
94. D.
95. A.
96. B.
97. C.
98. D.
99. A.
100. B.
101. C.
102. D.
103. A.
104. B.
105. C.
106. D.
107. A.
108. B.
109. C.
110. D.
111. A.
112. A.
113. B.
114. C.
115. D.
116. A.
117. B.
118. C.
119. D.
120. A.
121. A.
122. B.
123. C.
124. D.
125. A.
126. B.
127. C.
128. D.
129. A.
130. B.
131. C.
132. A.
133. A.
134. B.
135. C.
136. D.
137. A.
138. B.
139. C.
140. D.
141. A.
142. A.
143. B.
144. C.
145. D.
146. A.
147. B.
148. C.
149. D.
150. A.
151. A.
152. D.
153. C.
154. B.
155. C.
156. C.
157. D.
158. A.
159. D.
160. B.
161. C.
162. A.
163. A.
164. B.
165. C.
166. B.
167. B.
168. A.
169. C.
170. D.
171. D.
172. A.
173. B.
174. B.
175. C.
176. D.
177. C.
178. A.
179. D.
180. B.
181. C.
182. C.
183. B.
184. D.
185. D.
186. A.
187. A.
188. A.
189. C.
190. D.
191. B.
192. A.
193. B.
194. C.
195. D.
196. A.
197. B.
198. A.
199. A.
200. B.
201. A.
202. B.
203. B.
204. D.
205. D.

1. C. The development of the project charter is part of the project initiation process and is carried out through the Develop Project Charter process. The other tasks are performed outside the Initiating process group after the development of the project charter.

   Reference: *PMBOK® Guide*, 4th Ed., page 71

2. B. The project charter is the document that formally authorizes a project. The project charter provides the project manager with the authority to apply organizational resources to project activities. A project manager is identified and assigned as early in the project as is feasible. The project manager should always be assigned prior to the start of planning, and preferably while the project charter is being developed. Without approval of the charter, the project manager has no authority to assign resources to the project.

   Reference: *PMBOK® Guide*, 4th Ed., page 73

3. C. The project charter inputs may include the contract if the project is being done for an external customer. The other options are all subsets of an agreed-upon contract or calculations that lead to an agreed-upon contract.

   Reference: *PMBOK® Guide*, 4th Ed., page 75

4. D. The project statement of work (SOW) is provided by the project initiator or sponsor for internal projects based on a business need, product, or service requirement.

   Reference: *PMBOK® Guide*, 4th Ed., page 75

5. D. As a project manager, you must consider many factors such as the standards that are relevant to the project, the current and planned infrastructure of the organization, and the prevailing market conditions when developing a project charter. The other options are examples of these factors except for the introduction of a new course at another university.

   Reference: *PMBOK® Guide*, 4th Ed., page 76

6. B. When developing a project budget, a project manager may need to use a number of tools and techniques of economic analysis on the project benefits. One common method, especially in a capital project, is to calculate the payback period to identify where on the project schedule income exceeds the outgoing plus the initial investment. For this project, it will take three years for the Cash In (0 + 250,000 + 450,000) to equal the Cash Out (450,000 + 200,000 + 50,000).

   Reference: *PMBOK® Guide, 4th Ed.*, pages 77, 168

7. C. As a project manager you must not start a project without an approved charter. If the project starts without approval, organizational resources may be misdirected or wasted, and rework may be created. Your authority to proceed should be given by the ultimate authority, the project sponsor.

   Reference: *PMBOK® Guide*, 4th Ed., page 73

8. D. As a project manager, you must adopt or adapt the project management methods that exist to help you develop the project charter. Organizational process assets include any or all process-related assets and can include formal and informal plans, policies, procedures, and guidelines. A company's project management methodology falls into this category.

Reference: *PMBOK® Guide*, 4th Ed., pages 32, 76

9. D. When developing a project charter, the input of expert judgment often is used to help identify the inputs that must be considered in carrying out the Develop Project Charter process.

Reference: *PMBOK® Guide*, 4th Ed., page 77

10. B. When you are planning and implementing a project, one of your considerations is the project environment, including the cultural and social issues that may impact the success of the project. All of the other choices are general organizational or detail items used or evaluated later in the planning processes.

Reference: *PMBOK® Guide*, 4th Ed., page 14

11. C. The business need for the project is an essential input to the project charter, which is often stated through a business case. The detailed estimates and lists of risks are produced as part of project planning, which comes after the project chartering process. Some general and top-level resources are included in the project charter, but not all of the resources for the project are included.

Reference: *PMBOK® Guide*, 4th Ed., page 75

12. B. The WBS is a deliverable-oriented hierarchical decomposition of the project work. It is created through the Create WBS process and defines the total scope of the project.

Reference: *PMBOK® Guide*, 4th Ed., page 116

13. B. Configuration verification and audit ensures that the composition of a project's configuration items is correct. This activity also confirms that corresponding changes are registered, approved, tracked, and correctly implemented, to ensure that the functional requirements defined in the configuration document have been met. Option A refers to scope verification, option C is the definition of quality control, and option D describes monitoring and controlling project work.

Reference: *PMBOK® Guide*, 4th Ed., page 95

14. B. The project sponsors or initiators have the business need for the project. They approve the project scope statement for the project through the approval of a scope baseline. The scope baseline contains the project scope statement, WBS, and WBS dictionary. The roles in the other answers have input in creating the project scope statement but do not approve it.

Reference: *PMBOK® Guide*, 4th Ed., page 77

15. C. The business case contains the business need and cost-benefit analysis that justifies the go-ahead of the project, and it is created as a result of market demand, organizational need, customer request, technological advance, or legal requirement. All other options refer to other inputs used within the Develop Project Charter process.

Reference: *PMBOK® Guide*, 4th Ed., pages 75–76

16. B. Change requests can include corrective actions, preventive actions, and defect repairs. *Situational action* is a made-up term.

Reference: *PMBOK® Guide*, 4th Ed., page 92

17. B. The project management team is responsible for taking the project charter and creating a project scope statement. They are considered key stakeholders. It is important that all stakeholders provide input to the project scope statement.

Reference: *PMBOK® Guide*, 4th Ed., page 112

18. B. The WBS does not show responsibilities for tasks; the business need is defined in the project charter and dates are decided based on more detailed schedule planning. The WBS serves as a communication tool by defining 100 percent of the project scope.

Reference: *PMBOK® Guide*, 4th Ed., page 116

19. C. Approved project scope defines the approved work that the project delivers. The Control Scope process is responsible for ensuring that only the approved work is completed—no more, no less. Any changes made to the scope are managed by reference to the scope baseline, which represents the approved and signed-off scope.

Reference: *PMBOK® Guide*, 4th Ed., page 103

20. D. In a functional organization, the project manager has little or no influence on project outcomes. The functional manager controls the budget and the project manager is generally part-time, with part-time staff.

Reference: *PMBOK® Guide*, 4th Ed., page 28

21. B. The requirements traceability matrix ensures that requirements approved in the requirements documentation are delivered at the end of the project. The requirements traceability matrix also ensures that requirements add value by linking them to the business need they are addressing. Option A describes the project scope statement, option C is the project statement of work used in developing the project charter, and option D describes the business case.

Reference: *PMBOK® Guide*, 4th Ed., page 111

22. C. The use of expert judgment is recommended when creating the project scope statement. Experienced managers is not a clear answer. Special interest groups are a possibility, as are similar project plans, but expert judgment is considered the best answer because of the technical detail that can be provided by a subject matter expert.

Reference: *PMBOK® Guide*, 4th Ed., page 114

23. B. The project charter does not give the details of how the project will be executed, monitored, controlled, and closed. Process improvement and scope management plans are components of the project management plan.

Reference: *PMBOK® Guide*, 4th Ed., page 78

24. C. Change control is the process of identifying, requesting, assessing, validating, and communicating changes to the project management plan. It does not include measuring and does not include proving.

Reference: *PMBOK® Guide*, 4th Ed., page 94

25. D. The project manager, in conjunction with the project management team, directs, manages, and executes the project. The project sponsor or initiator does not manage the day-to-day activities of the project. The business unit/customer receives the deliverables from the project. The project manager cannot do it all. She needs to work with the team to execute the range of project activities to be performed.

Reference: *PMBOK® Guide*, 4th Ed., page 83

26. A. The project management plan is the main reference for the project manager during the execution stage of a project. The procurement management plan, communication management plan, and scope management plan are all subsets of the project management plan.

Reference: *PMBOK® Guide*, 4th Ed., pages 81–83

27. B. The definition of the execution stage is obtaining, managing, and using resources to achieve the project objectives. The other stages do not match the description given of the activities.

Reference: *PMBOK® Guide*, 4th Ed., page 83

28. C. The Direct and Manage Project Execution process has a number of inputs defined. All of these are approved documents or actions. The correct option is approved change requests. The other options are not inputs to the execution process.

Reference: *PMBOK® Guide*, 4th Ed., page 85

29. D. Only approved change requests should be scheduled into the project activities. Likely changes have not been approved and should not be scheduled into the project workload. Requests for change are approved only through the formal change control process.

Reference: *PMBOK® Guide*, 4th Ed., page 85

30. C. Deliverables are produced as outputs from the processes described in the project management plan. Not all tasks produce project deliverables.

Reference: *PMBOK® Guide*, 4th Ed., page 57

31. D. Project deliverables are outputs from the execution stage and are defined as a result, product, or capability produced and provided to complete the project. Options B and C are items obtained to complete the project. All items consumed in the project during the execution are consumables of the project.

Reference: *PMBOK® Guide*, 4th Ed., page 87

32. B. Option A does not give progress information. Option C does not give start or variance information. Option D is possible but not advisable, and it does not allow for short activities (less than one month in duration). Work performance measurement should be done routinely and regularly as part of the execution of the project management plan.

Reference: *PMBOK® Guide*, 4th Ed., page 89

33. C. Comparing actual activity performance against the project management plan identifies deviations and problems early. Broadcasting the plan does not measure progress. Recording progress does not refer to the plan or commitment of the project. Completed activity does not refer to the plan.

Reference: *PMBOK® Guide*, 4th Ed., page 89

34. D. The earned value management technique is used to predict future performance based on past performance. Future performance cannot be manipulated to exactly match the plan, past performance can be measured only as closely as actual data allows, and earned value cannot predict absolutely to 10 percent.

Reference: *PMBOK® Guide*, 4th Ed., pages 181–182

35. A. Preventive actions are documented actions that aim to reduce the probability of negative consequences associated with project risks. Increasing the project budget to allow for some overrun on costs or increasing the reporting frequency on activities that are critical could be considered preventive actions.

Reference: *PMBOK® Guide*, 4th Ed., page 92

36. B. In the Monitoring and Controlling process group, the Perform Integrated Change Control process applies from the beginning to the end of the project. The use of change control at one stage only is not recommended because many items may change and could be missed at other stages of the project.

Reference: *PMBOK® Guide*, 4th Ed., page 93

37. C. An integrated change control system must have a review step that results in either approval or rejection of changes to the project. The review step is applied not only to budget or to schedule change requests, but also to other change requests that can affect the project scope.

Reference: *PMBOK® Guide*, 4th Ed., page 93

38. D. The Perform Integrated Change Control process is part of the project management methodology. It aids the project management team in managing changes to the project.

Reference: *PMBOK® Guide*, 4th Ed., page 97

39. A. The Perform Integrated Change Control process has a number of outputs. Among them are project management plan updates. The project management plan is an input to this process, and is used by the change control board in making decisions to accept or reject changes.

Reference: *PMBOK® Guide*, 4th Ed., pages 98–99

40. B. The PMO is a central function for the coordination of the management of projects in an organization. The other options are not correct, and the project manager is responsible for closing out his own respective projects.

Reference: *PMBOK® Guide*, 4th Ed., page 11

41. C. The Close Project or Phase process is defined within the project management plan. The procurement, scope, and quality management plans are specific plans for each of these areas and do not specify how the project or phase should be closed.

Reference: *PMBOK® Guide*, 4th Ed., page 99

42. D. The Close Project or Phase process has accepted deliverables as an input. The final product/service, project management plan updates, and change request status updates are outputs of other processes.

Reference: *PMBOK® Guide*, 4th Ed., pages 101–102

43. C. The Delphi technique uses anonymous expert judgment and feedback to help bring consensus. Observation is the process of directly viewing individuals and how they perform tasks. Facilitated workshops are sessions that bring key cross-functional stakeholders together. Monte Carlo is a computer simulation for what-if scenario analyses.

Reference: *PMBOK® Guide*, 4th Ed., page 108

44. B. The requirements management plan documents how requirements will be analyzed, documented, and managed throughout the project. Option A is a definition of the requirements traceability matrix, option C describes the requirements documentation, and option D describes the scope management plan.

Reference: *PMBOK® Guide*, 4th Ed., page 110

45. C. The Close Project or Phase process involves administrative closure. This includes methodologies for how the project will transfer the services produced to the operational division of the organization. Contract closure is not concerned with transfer to operations. The project management plan is the superset of these procedures. Deliverable acceptance is not part of the Close Project or Phase process.

Reference: *PMBOK® Guide*, 4th Ed., page 100

46. C. The project closure documents include sponsor/customer acceptance documentation from scope verification. Administrative closure, contract closure, and organizational process assets do not require direct involvement of the sponsor.

Reference: *PMBOK® Guide*, 4th Ed., page 102

47. A. As part of the Close Project or Phase process, the project manager is responsible for collecting all the documentation resulting from the project activities. These include the project management plan, which itself contains the project's baselines, project calendar, and risk register, among others, which are known collectively as project files. The other options refer to items that are part of the organizational process assets, which project files will later become a part of.

Reference: *PMBOK® Guide*, 4th Ed., page 102

48. B. Project closure documents are formal documents that indicate the completion or termination of a project. Contract closure and administrative closure are separate activities.

Reference: *PMBOK® Guide*, 4th Ed., page 102

49. C. The Project Scope Management knowledge area defines and controls what is and what is not included in the project. Project documentation management and project change control are procedures that allow the control of changes to the project.

Reference: *PMBOK® Guide*, 4th Ed., pages 103–104

50. D. The Create WBS process subdivides the major project deliverables into smaller, more manageable components. The WBS does not concern project duration. Determining the total work package count is not a function of the Create WBS process. The allocation of responsibilities is not a function of the WBS, although it may be included as part of the WBS dictionary.

Reference: *PMBOK® Guide*, 4th Ed., page 116

51. A. The scope management plan provides guidance on how project scope will be defined, documented, verified, managed, and controlled. The scope management plan may be formal or informal, highly detailed, or broadly framed, based on the project needs. Option B describes the Collect Requirements process, option C describes the project scope statement, and option D describes the WBS.

Reference: *PMBOK® Guide*, 4th Ed., page 104

52. A. The detailed project scope statement includes product scope description, product acceptance criteria, project deliverables, exclusions, constraints, and assumptions. Deferred change requests could be approved at some time in the future but have not yet been approved and do not authorize the work that implies a change in scope.

Reference: *PMBOK® Guide*, 4th Ed., page 115

53. B. Prototypes, not requirements documentation, serve as a working model of the expected product that the project will deliver prior to the actual build. The rest of the options are valid requirements documentation descriptions.

Reference: *PMBOK® Guide*, 4th Ed., pages 109–110

54. C. A work breakdown structure template is often a WBS from a previous similar project within your organization because projects within an organization often share many common elements and processes. Generic or general-purpose lists of elements or structures are not always necessarily the best fit for your organization.

Reference: *PMBOK® Guide*, 4th Ed., pages 116–117

55. D. The work to be done in a project is defined in the WBS and the approved project scope statement. These two documents, plus the WBS dictionary, make up what is known as the scope baseline. Approved changes are only one part of the scope. Newly identified risks are not necessarily translated to impact on the work or scope. The WBS template is a generic starting point for defining the scope or final WBS.

Reference: *PMBOK® Guide*, 4th Ed., page 122

56. A. Verify Scope is the process of obtaining stakeholders' formal acceptance of the project and associated deliverables. This process is performed at the end of a phase, or before project closure, to determine if the project has delivered the contracted scope. Define Scope is the process of defining what the project will deliver. Control Scope is the process of managing changes to the scope. Scope Management refers to the overall scope-related efforts, defined in the project management plan, of how the team will define and manage scope.

Reference: *PMBOK® Guide*, 4th Ed., page 123

57. B. The act of monitoring and controlling product and project scope, as well as ensuring that all scope changes go through integrated change control, is part of the Control Scope process. The remaining options are incorrect because they are all part of the Planning process group.

Reference: *PMBOK® Guide*, 4th Ed., page 125

58. C. The Verify Scope process uses a number of techniques, including inspection. Inspection is a process of measuring, examining, and verifying that the work and deliverables meet the product and acceptance criteria. This is not a definition of Define Scope process inputs. Verify Scope outputs are not as described, and they include accepted deliverables. Scope management inputs are documents such as a scope statement and a list of deliverables.

Reference: *PMBOK® Guide*, 4th Ed., page 124

59. D. Along with the Verify Scope process, the Control Scope process is part of the Project Scope Management knowledge area. Specifically, these processes occur as part of the Monitoring and Controlling process group.

Reference: *PMBOK® Guide*, 4th Ed., pages 43, 125

60. A. Option A, configuration control and change control, respectively, is correct. As the name implies, change requests are modifications to agreed-upon requirements.

Reference: *PMBOK® Guide*, 4th Ed., page 94

61. A. Changes to project scope or requirements must be reviewed by the change control board through the Perform Integrated Change Control process. The change control board will review and then approve or reject the changes according to the documented guidelines. No work should start until it is approved by the change control board. Ignoring the change requests can lessen the probability of success for the project.

Reference: *PMBOK® Guide*, 4th Ed., page 125

62. B. Updating the project management plan is an output of the Control Scope process, once the changes have been approved. Some assignments may have to be changed as a result. Looking for other changes without issuing the new plan may cause rework. Letting the project schedule continue as previously and ignoring the new changes may compromise the project.

Reference: *PMBOK® Guide*, 4th Ed., page 128

63. C. Identifying the activities, dependencies, and resources needed to produce the project deliverables are some of the actions carried out in the Project Time Management knowledge area. Project Risk Management deals with risks and not with identifying and sequencing activities. Project Schedule Control and Project Cost Planning are not knowledge areas.

Reference: *PMBOK® Guide*, 4th Ed., pages 129–132

64. D. As part of the Project Time Management knowledge area, the project manager must define activities, which allows for decomposing the work packages into smaller components of work called *activities*. These activities are to be the basis for estimating, scheduling, and performing work of the project. Scheduling the work is part of the follow-on process, as are estimating tasks and scheduling activities.

Reference: *PMBOK® Guide*, 4th Ed., pages 133–136

65. A. The activity list is a comprehensive list of all scheduled activities, including the activities' scope and unique identifier that is in sufficient detail for a team member to complete the work. Defining the project scope is an earlier stage and does not break the project into detailed work packages. Defining the project schedule is a later stage of the planning. The WBS elements already will have been identified prior to this activity.

Reference: *PMBOK® Guide*, 4th Ed., page 135

66. B. A diagram on which activities are represented as boxes and the dependencies between activities are shown as arrows is known as activity on node, also known as a precedence diagram. Activity on arrow and schedule analysis are defined differently. The critical path method is a specific schedule network analysis technique.

Reference: *PMBOK® Guide*, 4th Ed., pages 138–139

67. C. The type of dependency in which the successor activity does not start until the completion of the predecessor activity is known as finish-to-start.

Reference: *PMBOK® Guide*, 4th Ed., page 138

68. D. A list that identifies and describes the types and quantities of resources required to complete activities is known as activity resource requirements. The resource calendar, project document updates, and resource breakdown structure are other outputs of the Estimate Activity Resources process.

Reference: *PMBOK® Guide*, 4th Ed., page 145

69. A. Estimating the duration of a project activity by referring to the actual duration of a similar activity on another project is known as analogous estimating. Expert judgment relies on specialists in a specific domain to perform a specialized task. Parametric estimating uses a metric from the current project in combination with historical information. Reserve analysis refers to determining the need for contingency, also known as time reserves or buffers.

Reference: *PMBOK® Guide*, 4th Ed., page 149

70. B. Calculating the theoretical early start and finish dates, along with late start and finish dates, for all the project activities using a forward and backward pass analysis is known as the critical path method. The critical chain method takes into account resource constraints. Schedule compression is about reducing project duration. Within the context project schedule, what-if scenario analysis attempts to answer the question: "If Scenario X occurs, what will be the impact to the project?"

Reference: *PMBOK® Guide*, 4th Ed., pages 154–155

71. C. A project overview that shows only the start and end of major deliverables, along with key external dependencies, is known as a milestone chart. Network diagrams and summary bar charts are used for more detailed presentation, and they also show dependencies and durations of tasks. The schedule baseline refers to the approved and signed-off version of the schedule.

Reference: *PMBOK® Guide*, 4th Ed., pages 157–159

72. D. A summary schedule lists the activity identifier, activity description, and calendar units with the length of the duration displayed proportionally as a horizontal bar relative to the project schedule time frame's start and end dates. The network diagram shows dependencies between activities. The detailed schedule with logical relationships shows more details than the summary schedule by displaying milestones and dependencies. The milestone schedule shows the calendar units as zero and only displays the milestones (end dates).

Reference: *PMBOK® Guide*, 4th Ed., page 158

73. A. The schedule baseline is the approved and signed-off version of the schedule.

Reference: *PMBOK® Guide*, 4th Ed., page 159

74. B. The Develop Schedule process uses the following tools and techniques: schedule network analysis, critical path method, critical chain method, resource leveling, what-if scenario analysis, use of leads and lags, schedule compression, and scheduling tool. The outputs include the project schedule, schedule baseline, schedule data, and project document updates.

Reference: *PMBOK® Guide*, 4th Ed., pages 154–159

75. C. Gathering actual start and finish dates of activities, along with remaining durations for work in progress, is known as a performance review. Cost variance analysis compares baseline to actual data. Performance measurement is a specific earned value technique. Critical path analysis is a planning tool.

Reference: *PMBOK® Guide*, 4th Ed., page 162

76. D. Variance analysis is used as part of the Control Schedule process to assess the magnitude of variation from the project schedule and schedule baseline in order to evaluate progress. Recording dates does not compare plan to actual. Calendar contract dates do not directly give variance analysis. The difference between total and free slack is not the variance that is analyzed in variance analysis.

Reference: *PMBOK® Guide*, 4th Ed., page 162

77. A. Schedule variance analysis, along with performance reviews, can result in change requests to the schedule baseline. *Action change control* is a made-up term and does not of itself constitute corrective actions. Updates to the schedule baseline may not be a result of the activity described. Updates to the project scope are not necessarily implied.

Reference: *PMBOK® Guide*, 4th Ed., page 164

78. B. Approved changes to the start and finish dates of the schedule baseline for the project are referred to as schedule baseline updates. Variance analysis is a technique that compares planned to actual dates, and a performance review involves reviewing, measuring, and analyzing schedule performance. The description is not a list of approved change requests.

Reference: *PMBOK® Guide*, 4th Ed., page 164

79. C. Estimating, budgeting, and cost control processes make up the Project Cost Management knowledge area. Project Risk Management is focused on risks, not solely on costs. Schedule Management concerns the duration and dependencies of tasks. Project Human Resource Management involves identifying necessary resources and checking that they are available.

Reference: *PMBOK® Guide*, 4th Ed., pages 165–168

80. B. As more information becomes known later in the project, estimates could narrow to a range of +/–10 percent. Earlier in a project, for example in the initiation stage, estimates could use a rough order of magnitude (ROM) range of +/–50 percent.

Reference: *PMBOK® Guide*, 4th Ed., page 168

81. A. Using only costs from a previous similar project that has been completed in your organization is known as analogous estimating. Bottom-up estimating aggregates the individual detailed estimates to arrive at the total estimate, whereas parametric estimating relies on statistical relationships to arrive at an estimate based on historical information and other parameters (for example, length, width, square footage, and so forth). As the name implies, three-point estimating uses three estimates to calculate the expected duration—most likely, optimistic and pessimistic.

Reference: *PMBOK® Guide*, 4th Ed., pages 171–172

82. C. Parametric estimating relies on statistical relationships to arrive at an estimate based on historical information and other parameters (for example, length, width, square footage, and so forth), whereas bottom-up estimating aggregates the individual detailed estimates to arrive at the total estimate. Using only costs from a previous similar project that has been completed in your organization is known as analogous estimating. As the name implies, three-point estimating uses three estimates to calculate the expected duration—most likely, optimistic and pessimistic.

Reference: *PMBOK® Guide*, 4th Ed., page 172.

83. C. The ETC requires the budget at completion (BAC), earned value (EV), and actual costs (AC) to date.

Reference: *PMBOK® Guide*, 4th Ed., page 185

84. B. The SPI is the only indicator that would help to measure schedule progress. The others are related only to costs.

Reference: *PMBOK® Guide*, 4th Ed., page 183

85. C. Earned value is used as part of the calculations for CV (EV – AC), SV (EV – PV), CPI (EV / AC) and SPI (EV / PV).

Reference: *PMBOK® Guide*, 4th Ed., pages 181, 183

86. A. Monitoring and recording results of executing quality activities to assess performance and recommend necessary changes occurs as part of the Perform Quality Control process. Perform Quality Assurance is responsible for ensuring that appropriate quality standards are used and for carrying out process improvement. Plan Quality involves establishing the requirements and/or standards for the project. Quality Improvement deals with increasing the effectiveness and efficiencies of organizational policies, processes, and procedures.

Reference: *PMBOK® Guide*, 4th Ed., pages 189–191

87. C. Plan Quality is related to establishing the quality requirements and/or standards for the project. Monitoring and recording results of executing quality activities to assess performance and recommend necessary changes is carried out through the Perform Quality Control process. Perform Quality Assurance is responsible for ensuring that appropriate quality standards are used, and for carrying out process improvement. Quality Improvement is an organizational development process.

Reference: *PMBOK® Guide*, 4th Ed., page 192

88. B. Perform Quality Assurance involves ensuring that appropriate quality standards are used and carrying out process improvement. Plan Quality is related to establishing the quality requirements and/or standards for the project. Monitoring and recording results of executing quality activities to assess performance and recommend necessary changes is a part of Perform Quality Control. Quality Improvement deals with increasing the effectiveness and efficiencies of organizational policies, processes, and procedures.

Reference: *PMBOK® Guide*, 4th Ed., page 201

89. D. A plan that details the steps for analyzing processes to identify activities that enhance the value of those processes in a project is known as the Perform Process Improvement process. The Perform Quality Assurance process includes auditing the quality requirements and assessing the results of quality measurements. Quality checklists are structured tools used to ensure that a predefined set of steps has been followed to ensure consistent delivery of a particular component. Quality metrics define the quality control measurements.

Reference: *PMBOK® Guide*, 4th Ed., page 201

90. A. Quality metrics clarify what some of the measurements on the project are and how the Perform Quality Control process will measure those metrics. Quality checklists are structured tools used to ensure that a predefined set of steps have been followed to ensure consistent delivery of a particular component. The quality management plan describes how the project will adhere to the organization's quality policy. A plan that details the steps for analyzing processes to identify activities that enhance the value in a project is known as the process improvement plan.

Reference: *PMBOK® Guide*, 4th Ed., page 200.

91. A. Quality metrics are an input to the Perform Quality Assurance process. Project requirements are not quality inputs. Change requests are an output of Perform Quality Assurance. Validated deliverables are an output of Perform Quality Control.

Reference: *PMBOK® Guide*, 4th Ed., page 203

92. B. Identifying the good/best practices being implemented, identifying gaps and shortcomings, sharing the good practices introduced or implemented in similar projects in the organization and/or industry, proactively improving process implementation, and highlighting the results in the Lessons Learned repository is known as performing a quality audit. Quality metrics are an input to Perform Quality Assurance and define the quality control measurements. Quality improvement deals with increasing the effectiveness and efficiencies of organizational policies, processes, and procedures. Quality checklists are structured tools used to ensure that a predefined set of steps has been followed to ensure consistent delivery of a particular component.

Reference: *PMBOK® Guide*, 4th Ed., page 204

93. C. The technique used to carry out the steps outlined in the process improvement plan to improve processes and project performance is known as process analysis. Quality metrics are an input to Perform Quality Assurance and define the quality control measurements. Performing a quality audit involves identifying inefficient and ineffective policies, processes, and procedures, and auditing results of the Perform Quality Control process. Quality checklists are structured tools used to ensure that a predefined set of steps has been followed to ensure consistent delivery of a particular component.

Reference: *PMBOK® Guide*, 4th Ed., page 204

94. D. Carrying out the Perform Quality Control process should occur throughout the project.

Reference: *PMBOK® Guide*, 4th Ed., page 206

95. A. Actions that have been recommended to increase the effectiveness and efficiency of the organization are documented as change requests. Change requests can be used to take corrective or preventive action or to perform defect repair. Organizational process assets updates typically involve updates to quality standards. Project management plan updates are related to changes to the quality, schedule, or cost management plans.

Reference: *PMBOK® Guide*, 4th Ed., page 205

96. B. A diagram that shows how various factors might be linked to the problem or the effects is known as a cause and effect diagram. A variable scatter diagram shows the relationship between two variables only. A process control chart is used to determine the stability of a system. Statistical sampling is related to population sampling of components.

Reference: *PMBOK® Guide*, 4th Ed., pages 208–209

97. C. A chart that shows, by frequency of occurrence, events and defects in the project is known as a Pareto chart. A Pareto chart is a type of histogram that incorporates the 80/20 principle. A scatter diagram shows the relationship between two variables only. A flowchart is a graphical representation of a process. A control chart relates to stability of processes.

Reference: *PMBOK® Guide*, 4th Ed., pages 210–211

98. D. A scatter diagram shows the relationship between two variables. A Control chart relates to the stability of processes. A Pareto chart shows, by frequency of occurrence, events and defects in the project. A run chart shows the history of the occurrence of a variation in a process.

Reference: *PMBOK® Guide*, 4th Ed., page 212

99. A. Identifying and documenting project roles, responsibilities, and reporting relationships, as well as creating a staffing management plan for a project, is carried out through the Develop Human Resource Plan process. The Develop Project Management Plan is a process that ensures that all the subsidiary plans are coordinated and integrated properly. To ensure an optimal utilization of human resources, the Manage Project Team process deals with project team members, from tracking performance to providing feedback, and from resolving issues to managing changes. The Develop Project Team process focuses primarily on enhancing project performance by providing proper training, facilitating team interaction, and creating a suitable work environment.

Reference: *PMBOK® Guide*, 4th Ed., pages 48, 58, 215–216

100. B. A matrix that illustrates the link between work packages and project team members is called a responsibility assignment matrix (RAM). A hierarchy-type chart resembles a traditional organizational chart by showing positions and relationships in a top-down format. Organizational charts can take many forms, such as hierarchical, matrix, or text-oriented. Network diagrams show the logical workflow of activities in a project.

Reference: *PMBOK® Guide*, 4th Ed., pages 220–221

101. C. A chart that illustrates the number of hours that each person will be needed each week over the course of the project schedule is known as a resource histogram. A WBS does not show weekly allocation. A task network diagram is not used to show schedule information for resources. A Gantt chart shows schedule activity information.

Reference: *PMBOK® Guide*, 4th Ed., page 224

102. D. Balancing out the peaks and valleys of resource usage is a commonly used resource-leveling strategy. None of the other strategies listed is a resource-leveling strategy.

Reference: *PMBOK® Guide*, 4th Ed., page 224

103. A. Enterprise environmental factors are an input of the Acquire Project Team process. As part of this input, the following is considered: availability, competencies, experience, interests, and costs of the potential project team. The project resource histogram shows use of resources on a time base. The responsibility assignment matrix shows which project team member should be working on a particular work package or activity. The organizational breakdown structure does not include the topics in the question.

Reference: *PMBOK® Guide*, 4th Ed., pages 221, 227

104. B. Defining project staff assignments within the project charter is known as preassignment. Resource management and resource leveling are activities not included in the project charter. The responsibility assignment matrix shows which project team member should be working on a particular work package or activity.

Reference: *PMBOK® Guide*, 4th Ed., pages 221, 227

105. C. A team whose members all have a shared goal and all perform their roles, spending little or no time meeting face to face, is known as a virtual team. Composite (a blend of functional and projectized) and matrix organizations are types of organizational structures. Remote teams do not necessarily have a shared goal.

Reference: *PMBOK® Guide*, 4th Ed., pages 31, 228

106. D. The scenario described acquisition, which is a tool or technique of the Acquire Project Team process. This question is slightly difficult to answer because you need to know the tool or technique first, and then you need to relate it to the correct process.

Reference: *PMBOK® Guide*, 4th Ed., pages 227–228

107. A. The Develop Project Team process includes activities such as improving the competencies of team members, improving team interaction, and improving the environment. Creating the resource calendars and assigning team members to the project are outputs of the Acquire Project Team process.

Reference: *PMBOK® Guide*, 4th Ed., page 229

108. B. The Develop Project Team process uses the following tools and techniques: interpersonal skills, training, team-building activities, ground rules, co-location, and recognition and awards.

Reference: *PMBOK® Guide*, 4th Ed., pages 229, 230

109. C. Understanding better the sentiments of the project team members and acknowledging their concerns are examples of interpersonal skills. Team building helps project team members to work more effectively. Ground rules encompass setting clear expectations within the project team. Co-location entails placing project team members in one physical location to enhance their ability to perform as a team.

Reference: *PMBOK® Guide*, 4th Ed., pages 232–234

110. D. Interpersonal skills, training, and team-building activities are tools and techniques of the Develop Project Team process. Conflict management is one of the tools and techniques in the Manage Project Team process.

Reference: *PMBOK® Guide*, 4th Ed., pages 232, 236

111. A. An offsite meeting designed to improve interpersonal relationships within the project team is an example of a team-building activity. Team building occurs as part of the Develop Project Team process. Soft Skills Training is a specific type of training related to interpersonal skills. Effective Team Working is related to plan activities in a project schedule designed to enhance the competencies of project team members. Improving Team Competencies is related to developing the project team.

Reference: *PMBOK® Guide*, 4th Ed., pages 232, 233

112. A. Only desirable behavior should be rewarded in a recognition and reward system. Rewarding the team member of the month and rewarding all team members who work overtime has a narrow focus compared to focusing only on desirable behavior. Rewarding individualism regardless of culture may not always be appropriate.

Reference: *PMBOK® Guide*, 4th Ed., page 234

113. B. Managing the project team becomes more complicated when team members are accountable to both a functional and a project manager, as is the case within a matrix organization. Functional and projectized organizations have clear accountability. Hierarchical organizations are the norm.

Reference: *PMBOK® Guide*, 4th Ed., page 240

114. C. Tracking team performance, providing feedback, resolving issues, and coordinating changes to enhance project performance are all part of the Manage Project Team process. The Develop Project Team process focuses on optimizing team performance. The Acquire Project Team process ensures that the project team is available to do the work. Lastly, the Develop Human Resource Plan process deals with defining roles and responsibilities.

Reference: *PMBOK® Guide*, 4th Ed., page 236

115. D. Checking skills improvements, recording current competencies, and monitoring reduction in team turnover rates are part of the Develop Project Team process. Project staff assignments are an input to developing the project team. Recognition and reward systems, as well as team-building activities, are tools and techniques of this process.

Reference: *PMBOK® Guide*, 4th Ed., pages 235–236

116. A. Schedule control, cost control, quality control, scope verification, and procurement audits are all examples of inputs to managing the project team known as project performance reports. Team performance assessment is related to the team members explicitly. The staffing management plan is a list or schedule of work for staff. Organizational process assets are policies, procedures, and systems for reward.

Reference: *PMBOK® Guide*, 4th Ed., page 238

117. B. The organizational process assets that can influence the Manage Project Team process can include certificates of appreciation, newsletters, websites, bonus structures, corporate apparel, and other organizational perquisites. Team performance assessment is an appraisal of how the team is performing as a whole. The staffing management plan documents how the team will be acquired and released, trained, compensated, and so forth. Documented in the human resource plan, staff roles and responsibilities are related to individuals, not the performance.

Reference: *PMBOK® Guide*, 4th Ed., page 238

118. C. A document that lists who is responsible for resolving a specific problem by a target date is known as a project issue log. A project risk register is a specific tool for identifying, tracking, and resolving risks. Change control is the process that supports changes to any item or process within the project. Performance reports are related to staff management.

Reference: *PMBOK® Guide*, 4th Ed., page 240

119. D. Identifying and determining the information needs of project stakeholders, making information available in a timely manner, collecting and distributing performance information, and meeting information needs to help resolve issues with stakeholders are examples of activities that occur as part of the Project Communications Management knowledge area.

Reference: *PMBOK® Guide*, 4th Ed., pages 243–245

120. A. Inputs to the Plan Communications process include stakeholder register, stakeholder management strategy, enterprise environmental factors, and organizational process assets. Tools and techniques consist of communication requirements analysis, communication technology, communication models, and communication methods. There are only two outputs, namely communications management plan and project document updates.

Reference: *PMBOK® Guide*, 4th Ed., page 252

121. A. Methods and technologies to be used to convey memoranda, email, and press releases, and how frequently these should be used, are examples of the contents of a communications management plan.

Reference: *PMBOK® Guide*, 4th Ed., pages 256–257

122. B. The staff member responsible for the distribution of information about the project is defined in the communications management plan. The project management plan is the higher-level document that contains the communication management plan within it. Project roles and responsibilities are related to task responsibilities. The staffing management plan lists the human resources needs of the project and how they will be met.

Reference: *PMBOK® Guide*, 4th Ed., pages 223, 256–257

123. C. Making information clear and complete so that the recipient can understand it correctly is an example of communications skills. Influence skills, negotiating skills, and delegation skills are different attributes of the soft skills set that a project manager should have.

Reference: *PMBOK® Guide*, 4th Ed., pages 260, 411

124. D. Written and oral communication techniques are a part of communications skills. Influence skills, negotiating skills, and delegation skills are different attributes of the soft skills set that a project manager should have.

Reference: *PMBOK® Guide*, 4th Ed., page 245

125. A. Identifying project successes and failures, and making recommendations on how to improve future performance on other projects, is part of the Lessons Learned documentation. The other choices are all inputs to the Manage Stakeholder Expectations process.

Reference: *PMBOK® Guide*, 4th Ed., page 261

126. B. Actions and suggestions that are documented and considered for use on future projects are called Lessons Learned.

Reference: *PMBOK® Guide*, 4th Ed., page 261

127. C. Generally regular reports are required by stakeholders, and exception reports should be issued as appropriate.

Reference: *PMBOK® Guide*, 4th Ed., page 266

128. D. Report Performance is the process of collecting and distributing performance information, including status reports, progress measurements, and forecasts. Baselining the project is a specific action to record the signed-off and approved versions of the schedule, budget, and scope. Stakeholder management refers to analyzing and communicating with the stakeholders. Team performance assessments are the output of the Develop Project Team process.

Reference: *PMBOK® Guide*, 4th Ed., pages 58, 266

129. A. Budget forecasts are an input to the Report Performance process. Performance reports, change requests, and organizational process assets updates are all outputs of this process.

Reference: *PMBOK® Guide*, 4th Ed., page 268

130. B. Work performance information on the status of deliverables and what has been accomplished in the project is part of the Executing process group. Project initiation deals with obtaining approvals to start a project or phase. Project planning establishes the project scope and objectives. Lastly, project closure formalizes the completion of the project or phase.

Reference: *PMBOK® Guide*, 4th Ed., pages 266–268

131. C. Information that includes bar charts, S-curves, histograms, and tables for the data analyzed against the performance measurement baseline is known as performance reports. Work performance measurements and budget forecasts are inputs to the Report Performance process.

Reference: *PMBOK® Guide*, 4th Ed., page 270

132. A. Performance reports are issued periodically and may range from simple status reports to more elaborate reports. More elaborate reports may contain analysis of past performance, forecasted project completion, current status of risks and issues, and results of variance analysis, among other information. Variance analysis, corrective actions, and change requests are not related to forecasting.

Reference: *PMBOK® Guide*, 4th Ed., page 270

133. A. Managing communications to satisfy the needs of, and resolve issues with, all the project stakeholders is usually the responsibility of the project manager. Team members, the project sponsor, and the communication specialist do all communicate with other stakeholders, but they are not responsible for managing the process.

Reference: *PMBOK® Guide*, 4th Ed., page 262

134. B. One of the tools used to help document and monitor the status and resolution of issues in a project is the issues log. The risk register captures the outcomes of the risk management processes. Change requests are actions required as a result of issues or risks identified. Corrective actions refer to changes made when executing the project.

Reference: *PMBOK® Guide*, 4th Ed., page 263

135. C. Identification, analysis, planning responses, and monitoring and controlling of risks in the project are part of the Project Risk Management knowledge area. Risk identification, analysis, and mitigation refer to various risk-related activities.

Reference: *PMBOK® Guide*, 4th Ed., page 273

136. D. Except for the Monitoring and Controlling Risks process, five of the six project risk management processes are part of the Planning process group.

Reference: *PMBOK® Guide*, 4th Ed., pages 43, 274

137. A. A balance between risk taking and risk avoidance refers to the application of risk responses, which are determined as part of the Plan Risk Responses process. Risk analysis, risk identification, and risk classification are other risk-related efforts that take place prior to determining the appropriate responses to identified risks.

Reference: *PMBOK® Guide*, 4th Ed., page 276

138. B. The probability and impact matrix is one of the tools used to manage risk in a project by showing the various thresholds for different levels of risks. A risk register records all the identified risks and information about them. An issues log records all issues. The impact scales are used to define the impact of risks relative to major project objectives (for example, cost, time, scope, and so forth).

Reference: *PMBOK® Guide*, 4th Ed., pages 281, 292

139. C. Any project personnel can identify risks in a project. The project manager or risk manager manages the risk management processes, with input from the sponsor and other stakeholders.

Reference: *PMBOK® Guide*, 4th Ed., page 282

140. D. Identified risks, risk responses, the root cause of the risk, and the risk category are the basic fields in a risk register. The risk management plan is the overall management document and processes definitions for managing risk in the project. The issues log contains the list of all issues. When a decision is made to transfer risk, risk-related contracts will include agreements to purchase insurance, bonds, and so forth in order to properly mitigate or transfer part or all of the risks.

Reference: *PMBOK® Guide*, 4th Ed., pages 288, 306

141. A. Rating risks as low, medium, or high is done using a probability impact matrix. Risk register updates are an output in five of the six risk management processes. Assumptions analysis and checklist analysis are techniques used for risk identification.

Reference: *PMBOK® Guide*, 4th Ed., pages 291–292

142. A. A technique that is often used for quantitative risk analysis that calculates or obtains information on the lowest, highest, and most likely costs of the WBS elements in the project plan is an example of three-point estimating for costs. These estimates can then be used for triangular, beta, or other distributions when performing cost risk simulations. Probability and impact assessment is a ranking of risks. Probability distributions, within the context of project risk management, are used to model and simulate uncertainties in the project, such as duration and cost estimates. Sensitivity analysis is used to test the major project variables independently for risks.

Reference: *PMBOK® Guide*, 4th Ed., pages 296–297

143. B. Developing options, determining actions to enhance opportunities, and reducing threats to project objectives is known as plan risk responses. Perform qualitative risk analysis and perform quantitative risk analysis are prior steps in the Risk Management knowledge area. The probability impact matrix is one of the tools used to manage risk in a project by showing the various thresholds for different levels of risks.

Reference: *PMBOK® Guide*, 4th Ed., page 301

144. C. The risk register is first created through the Identify Risks process. Qualitative and quantitative risk analysis follow this process. Plan Risk Responses is a process following these two earlier processes within the Project Risk Management knowledge area.

Reference: *PMBOK® Guide*, 4th Ed., page 288

145. D. A common risk management strategy used to shift the negative impact of a threat, and ownership of the response, to a third party is called transfer. Exploit is a risks response used for risk opportunities. Avoid and mitigate are alternative techniques for dealing with a negative risk.

Reference: *PMBOK® Guide*, 4th Ed., pages 303–305

146. A. For some of the risks in a project, particularly those that pose a threat as well as an opportunity, a response plan that will be executed only when certain predefined conditions exist is called a contingent response strategy. Sharing, exploit, and enhance are responses to opportunities, also known as positive risks.

Reference: *PMBOK® Guide*, 4th Ed., page 305

147. B. The project team should monitor the project work for new and changing risks continuously throughout the project lifecycle. The efforts carried out as part of the Monitor and Control Risk process should be defined in the risk management plan, and it should specify continuous monitoring for new and changing risks.

Reference: *PMBOK® Guide*, 4th Ed., page 308

148. C. Information such as identified risks, risk owners, agreed responses, actions, warning signs, and a watch list of low-priority risks are examples of fields from the risk register. The risk management plan defines the process and resources involved in managing the risks. Approved change requests and work performance information do not match the question items.

Reference: *PMBOK® Guide*, 4th Ed., page 309

149. D. The action of identifying and documenting the effectiveness of risk responses in dealing with identified risks and the root causes, and of the effectiveness of the risk management process, is known as carrying out a risk audit. Risk mitigation is a type of risk response. Risk analysis can be qualitative (prioritization based on probability and impact) or quantitative (numerically analyzing the effect of a risk relative to project objectives). Risk audits examine and document effectiveness of the Risk Management process.

Reference: *PMBOK® Guide*, 4th Ed., pages 310, 440

150. A. Actions that are required to bring the project into compliance with the project management plan are known as recommended preventive actions, whereas recommended corrective actions include contingency plans and workaround plans. Risk register updates and project management plan updates are also outputs of the Monitor and Control Risks process, but they do not fit the question posed.

Reference: *PMBOK® Guide*, 4th Ed., page 312

151. A. In a complex project involving many contracts and subcontractors, each contract lifecycle can end during any phase of the project, or when the deliverables are accepted.

Reference: *PMBOK® Guide*, 4th Ed., page 315

152. D. The process of deciding whether a particular product or service can be produced by the project team or instead should be purchased is known as make-or-buy analysis. Procurement management refers to the procurement-related activities and effort carried out throughout the project lifecycle. Expert judgment is a technique used in many decision-making processes. Risk management refers to all risk-related efforts carried out throughout the project's life in order to identify and manage risks.

Reference: *PMBOK® Guide*, 4th Ed., page 321

153. C. A contract that involves a fixed price for a well-defined requirement, possibly with incentives for meeting selected objectives, is a fixed-price contract. A time and material contract is a cross between a cost-reimbursable and a fixed-price contract. A cost-reimbursable contract includes payments for costs incurred to complete the work plus a fee to the seller. A cost plus incentive fee contract includes a predetermined incentive fee based on performance.

Reference: *PMBOK® Guide*, 4th Ed., pages 322–324

154. B. How well the seller meets the contract statement of work, whether the seller has the capacity to meet future requirements, and whether the seller can provide references from previous customers are examples of contract source selection criteria. Contract management is the overall effort of managing contracts. Make-or-buy decisions capture the conclusion of an evaluation whether a particular project component should be built by the project team or acquired externally. Contract negotiation occurs after source selection criteria have been decided and analyzed.

Reference: *PMBOK® Guide*, 4th Ed., pages 326–327

155. C. The procurement process involving obtaining seller responses, selecting a seller, and awarding a contract is called Conduct Procurements. Qualified sellers list is an input to this process, bidder conferences is a tool and technique used within this process, and select acceptable seller is an output of the process.

Reference: *PMBOK® Guide*, 4th Ed., pages 328–335

156. C. A list of prospective and previously qualified sellers and information on relevant past experience with sellers, both good and bad, is considered to be an organizational process asset. The qualified sellers list is one part of these assets. The procurement management plan and procurement documents are other inputs to the Conduct Procurements process.

Reference: *PMBOK® Guide*, 4th Ed., page 331

157. D. A structured meeting to ensure that all prospective sellers have a clear and common understanding of the procurement you require is called a bidder conference. Change control meetings are conducted to review whether to approve, defer, or reject proposed changes to the project. One-on-one meetings involve a conversation between two individuals. Project status meetings review the status of the project, including progress, issues, risks, costs, and so forth.

Reference: *PMBOK® Guide*, 4th Ed., page 331

158. A. A list of sellers who have been prescreened for their qualifications and past experience, so that procurements are sent only to those sellers who can perform for any resulting contract, is called a qualified sellers list. Procurement documents and organizational process assets, as well as this list, are inputs to the Conduct Procurements process. A procurement statement of work is an output from the Plan Procurements process.

Reference: *PMBOK® Guide*, 4th Ed., page 330

159. D. Seller proposals are prepared in response to a procurement document package, and they form the basic set of information that will be evaluated to select the successful bidder(s). Requirements are a list of the needs of the contract or project. Responses are the reply to request for tender. Requests are the initial requests from the customer.

Reference: *PMBOK® Guide*, 4th Ed., page 330

160. B. Establishing source selection criteria for each of your sellers' proposals helps to rate or score seller proposals. These source selection criteria are an output of the Plan Procurements process and are used as an input to the Conduct Procurements process. The bidder conference is an open meeting prior to the proposals being delivered, where the buyer answers questions from the potential sellers. The Administer Procurements process includes managing procurement-related contracts, relationships, and changes.

Reference: *PMBOK® Guide*, 4th Ed., pages 64, 327–328

161. C. The project management plan, seller proposals, and qualified seller list are examples of inputs to the Conduct Procurements process. They are not contract types.

Reference: *PMBOK® Guide*, 4th Ed., page 330

162. A. A response to a procurement document, such as a request for proposal (RFP), is called a proposal. While a presentation may be given by potential sellers as part of the proposal, it is not considered a formal response. Requests in this instance refer to the buyer's output. Bidding describes the response by potential sellers to a request by a buyer.

Reference: *PMBOK® Guide*, 4th Ed., page 330

163. A. Independent estimates are used to highlight significant differences between your pricing and the independent consultants' pricing, for example. Bidder conferences are used to ensure that all prospective sellers have a clear and common understanding of the goods or services being procured. Screening systems are used to establish minimum levels of compliance to requirements. Expert judgment is used to assist in the evaluation of sellers' proposals.

Reference: *PMBOK® Guide*, 4th Ed., page 332

164. B. A teaming agreement is a legal contractual agreement between two or more entities to form a partnership or joint venture, or some other arrangement as defined by the parties, and describes buyer-seller roles for each party. When the new business opportunity ends, so does the teaming agreement. A cost-reimbursable contract, a noncompete agreement, and a boilerplate agreement do not define this type of venture.

Reference: *PMBOK® Guide*, 4th Ed., page 319

165. C. The procurement contract is a mutually binding legal agreement that obligates the seller to provide the specified services and also obligates the buyer to pay the seller. A request, in relation to the question, is from the buyer, asking for proposals from sellers. The tender, a formal offer to produce a product or deliver a service, contains the documents provided by the seller. The proposal is the offering from the seller.

Reference: *PMBOK® Guide*, 4th Ed., page 333

166. B. Having proposals evaluated by a multidiscipline team including members with specialized knowledge in each of the areas covered by the proposed contract is known as using expert judgment. Seller rating systems are sometimes based on past performance. Screening systems are based on predefined minimum levels of compliance to requirements. Bidder conferences are used to ensure that all prospective sellers have a clear and common understanding of the procurement.

Reference: *PMBOK® Guide*, 4th Ed., page 332

167. B. The procurement management plan is a subset of the project management plan. Procurement documents are part of the project documents. Claims administration deals with the handling of claims, disputes, or appeals. The qualified sellers list is a separate document that is an input of the Conduct Procurements process.

Reference: *PMBOK® Guide*, 4th Ed., page 324

168. A. The process of ensuring that the seller's performance meets the contractual requirements and that the buying organization performs according to the contract is called Administer Procurements. Plan Procurements and Conduct Procurements are processes that lead up to the Administer Procurements process. Selected sellers is an output of the Conduct Procurements process.

Reference: *PMBOK® Guide*, 4th Ed., page 335

169. C. The procurement activities described within the procurement management plan are managed from the beginning to the end of the contract. The project lifecycle may be longer than the contract. Seller selection and developing contract documentation are activities that follow the creation of the procurement management plan.

Reference: *PMBOK® Guide*, 4th Ed., page 324

170. D. The contract with a seller can be modified at any time, in accordance with the change control terms of the contract and project. Contracts usually are varied or modified during a project, for practical reasons. The contract is not in place until it has been awarded.

Reference: *PMBOK® Guide*, 4th Ed., page 337

171. D. A contract change control system includes contract documentation, tracking systems, dispute resolution procedures, and approved levels of authority for changes. This is a subset of the change control system. The source selection criteria are an input to the Conduct Procurements process. The procurement management plan describes how the procurement processes will be managed.

Reference: *PMBOK® Guide*, 4th Ed., page 338

172. A. If the parties cannot resolve a claim by themselves, it may need to be handled in accordance with alternative dispute resolution, usually following procedures established in the contract. Compromising is a technique for conflict resolution within project teams; integrated change control is the process for handling changes, not claims; and economic price adjustment is a form of contract.

Reference: *PMBOK® Guide*, 4th Ed., page 339

173. B. A document that rates how well each seller is performing the project work is known as a seller performance evaluation. Source selection criteria are an input to Conduct Procurements. The procurement management plan refers to how the procurement processes will be managed. Work performance information is raw data that reflects the seller's performance.

Reference: *PMBOK® Guide*, 4th Ed., page 340

174. B. Verification that all the work and deliverables supplied by the seller were acceptable is carried out through the Close Procurements process. Close Project is not solely related to sellers. Lessons Learned are collected as part of performing project closure. Performance reporting is related to the seller's performance as the contract is in progress.

Reference: *PMBOK® Guide*, 4th Ed., page 341

175. C. Although all the solutions are possible, the consistent underperformance can help to indicate early termination of the seller contract.

Reference: *PMBOK® Guide*, 4th Ed., pages 341–342

176. D. Archiving the seller performance for future use is part of the Close Procurements process. Close Project or Phase refers to another closing process, and Lessons Learned are gathered as part of both closing processes. Performance reporting is an Administer Procurements process activity.

Reference: *PMBOK® Guide*, 4th Ed., pages 341–342

177. C. A structured review of the procurement processes from Plan Procurements through Administer Procurements is called a procurement audit. Contract management is the generic term for the overall process. Performance reporting and claims administration are other subsets of Contract Management.

Reference: *PMBOK® Guide*, 4th Ed., page 343

178. A. Procurement documentation is a key input to the Close Procurements process. Closed procurements, deliverable acceptance, and Lessons Learned are all outputs from the Close Procurements process.

Reference: *PMBOK® Guide*, 4th Ed., page 344

179. D. The objective of the use of the Procurement Audit process is to identify success and failure for use in future contracts on this or other projects. This is performed as part of the project closure process. Identifying when legal action should be started, terminating the nonperforming suppliers' contracts, and identifying who signed the nonperforming contracts are part of contract administration.

Reference: *PMBOK® Guide*, 4th Ed., page 343

180. B. Providing the seller of one of the project deliverables with formal notice that the contract has been completed is called Close Procurements. The Administer Procurements process includes managing procurement-related contracts, relationships, and changes. Lessons Learned and deliverable acceptance are organizational process assets.

Reference: *PMBOK® Guide*, 4th Ed., page 344

181. C. Providing the seller of one of the project deliverables with formal notice that the deliverable is acceptable is called deliverable acceptance. Procurement management and contract management are procurement-related efforts. Lessons Learned is an output of the Close Procurements process, through organizational process assets updates.

Reference: *PMBOK® Guide*, 4th Ed., page 344

182. C. A list of recommended process improvements for future contracts and purchasing is called Lessons Learned. Contract documentation is the generic term for all documents related to the contracts. Procurement audit and claims administration are parts of running a contract.

Reference: *PMBOK® Guide*, 4th Ed., page 344

183. B. Practitioners in the project management profession are committed to doing what is right and honorable as defined in the *Code of Ethics and Professional Conduct* vision statement. The other statements are details of applicability, structure, and values.

Reference: *Project Management Professional (PMP®) Handbook*, page 46

184. D. One of the purposes of the *Code of Ethics and Professional Conduct* in project management is to instill confidence in the project management profession. Keeping unqualified project managers out of contracts, enabling prosecution of poorly performing project managers, and identifying project managers who are not qualified are not the purposes of this code.

Reference: *Project Management Professional (PMP®) Handbook*, page 46

185. D. The basis used to develop the *Code of Ethics and Professional Conduct* were values that formed decision making and guided actions.

Reference: *Project Management Professional (PMP®) Handbook*, page 46

186. A. The standards of conduct are described as aspirational and mandatory. Aspirational and legal, mandatory and contractual. and aspirational and contractual do not refer to the *Code of Ethics and Professional Conduct.*

Reference: *Project Management Professional (PMP®) Handbook*, pages 47–49

187. A. Taking ownership for the decisions we make or fail to make, the actions we take or fail to take, and the resulting consequences is defined as the responsibility of project managers. Respectability is not as defined. Confidentiality and consistency are not as described and are not values as defined in the code.

Reference: *Project Management Professional (PMP®) Handbook*, page 46

188. A. Calling violations of the *Code of Ethics and Professional Conduct* to the attention of the appropriate body for resolution is part of the mandatory responsibility of the Project Management Institute. This is not an aspirational responsibility or aspirational respectability. *Optional responsibility* is not a defined term.

Reference: *Project Management Professional (PMP®) Handbook*, page 47

189. C. As project managers, showing high regard for ourselves, for others, and for the resources entrusted is respect. Honesty, responsibility, and fairness are the other three values described in the Project Management Institute's *Code of Ethics and Professional Conduct.*

Reference: *Project Management Professional (PMP®) Handbook*, pages 47–48

190. D. Taking time to listen to others' point of view, seeking to understand them, is aspirational respect.

Reference: *Project Management Professional (PMP®) Handbook*, page 48

191. B. Acting in a way that is abusive to another member of the team is a breach of the mandatory respect standard.

Reference: *Project Management Professional (PMP®) Handbook*, page 48

192. A. Making decisions and acting impartially and objectively is defined as the value of fairness.

Reference: *Project Management Professional (PMP®) Handbook*, page 48

193. B. Providing equal access to information to those who are authorized to have the information is an example of aspirational fairness.

Reference: *Project Management Professional (PMP®) Handbook*, page 48

194. C. Disclosure of potential conflict of interest is an example of mandatory fairness.

Reference: *Project Management Professional (PMP®) Handbook*, page 48

195. D. To try to understand the truth and act in a truthful manner in both communications and conduct is a definition of honesty.

Reference: *Project Management Professional (PMP®) Handbook*, page 40

196. A. Creating an environment in which others feel safe to tell the truth is an example of aspirational honesty.

Reference: *Project Management Professional (PMP®) Handbook*, page 49

197. B. Engaging in dishonest behavior for personal gain is a breach of the mandatory honesty code.

Reference: *Project Management Professional (PMP®) Handbook*, page 49

198. A. Conduct by a team member that results in others feeling humiliated is an example of abusive manner. This is not allowable behavior, and it is not permitted by any aspirational or mandatory standard in the Project Management Institute's *Code of Ethics and Professional Conduct.*

Reference: *Project Management Professional (PMP®) Handbook*, page 50

199. A. Violation of the Project Management Institute's *Code of Ethics and Professional Conduct* may result in sanctions by the Institute as defined in the ethics case procedures. There are no legal implications directly implied by the Code. Salary deductions are for the employer to consider. Name and shame may be done by the press.

Reference: *Project Management Professional (PMP®) Handbook*, page 50

200. B. A project manager's responsibility, both legally and morally, to promote the best interests of the organization is known as the duty of loyalty. This is good practice and may be an aspirational standard as well, but it is described explicitly as the duty of loyalty. It is not aspirational respect as defined in the *Code of Ethics and Professional Conduct.*

Reference: *Project Management Professional (PMP®) Handbook*, page 50

201. A. The most commonly used dependency in a PDM method is finish-to-start (FS). Start-to-finish (SF) means that a successor schedule activity depends on the initiation of the predecessor schedule activity. In finish-to-finish (FF), the successor schedule activity cannot finish until the completion of the predecessor schedule activity. In start-to-start (SS), the initiation of the successor schedule activity depends on the start of the predecessor schedule activity.

Reference: *PMBOK® Guide*, 4th Ed., page 138

202. B. The ADM, now rarely used along with its activity on arrow (AOA) because of computer-assisted scheduling, is visually the opposite of the precedence diagramming method (PDM). SPI is an acronym for schedule performance index and WBS is an acronym for work breakdown structure.

Reference: *PMBOK® Guide*, 4th Ed., pages 162, 166, and 353

203. B. Hours, days, and weeks are units of measure. All of the remaining options are incorrect.

Reference: *PMBOK® Guide*, 4th Ed., page 145

204. D. A communication model is a tool and technique of the Plan Communications process. Inputs to the Plan Communications process include stakeholder register, stakeholder management strategy, enterprise environmental factors, and organizational process assets.

Reference: *PMBOK® Guide*, 4th Ed., page 252

205. D. Avoid, transfer, mitigate, and accept are risk response strategies for negative risks. However, accept can also be used for positive risks.

Reference: *PMBOK® Guide*, 4th Ed., pages 303–305

Practice Test C

1. Managing projects involves a number of project management process groups that are required to accomplish the project objectives. Without proper integration management, a project is likely to fail. What is the primary focus of project integration management?

   A. Integrating the project plans with third-party plans

   B. Managing the integration of the customer business plan and the project plan

   C. Coordinating the various processes within the project management process groups

   D. Integrating the project plan with government legislative plans

2. A critical role for a project manager is to manage the project integration activities. Integration is primarily concerned with effectively integrating processes among the project management process groups that are required to accomplish project objectives within an organization's procedures. Which one of the following is *not* part of integrative project management processes?

   A. Perform Integrated Change Control

   B. Develop Project Charter

   C. Estimate Costs

   D. Close Project or Phase

3. A project should not start without an authorized project charter. Projects usually are chartered and authorized as a result of one or more stimulants, which can be problems, opportunities, or business requirements. Examples of valid stimuli for chartering projects by a business enterprise include all of the following *except* ______________.

   A. A social need

   B. A political conflict

   C. A legal requirement

   D. A technological advance

4. In most enterprises, a project is not formally chartered and initiated until the completion of a number of prerequisites. The following list gives examples of prerequisites for formally chartering and initiating a project *except* ______________.

   A. Business case

   B. Customer request

   C. Contract

   D. Project management methodology

5. The project charter is a mandatory document that is required before a project is allowed to start. The project charter, either directly or by reference to other documents, should address a number of information items. The following items are part of the contents of the project charter *except* ______________.

A. Project Purpose or Justification

B. Assigned Project Manager, Responsibility, and Authority Level

C. Plan Procurements

D. Measurable Project Objectives and Related Success Criteria

6. An important input to developing the project charter is the statement of work (SOW). The SOW is a narrative description of products or services to be supplied by the project. It indicates all of the following *except* ______________.

A. Project budget

B. Product scope description

C. Strategic plan

D. Business need

7. Most organizations have a knowledge base as part of their organizational process assets, which is available for all project management teams. When developing the project charter and subsequent project documentation, any or all of the assets that are used to influence the project's success can be drawn from organizational process assets. An important element of these assets is the organization's processes and procedures, which include all of the following *except* ______________.

A. Project Contracts

B. Project Closure Guidelines

C. Change Control Procedures

D. Work Instructions

8. Your project has been authorized by management, and you are in the process of planning. From the following list, which item is *not* an output of the project planning processes?

A. How work will be executed to achieve the project objectives

B. The project management processes selected by the project management team

C. The final products resulting from the project activities

D. The life cycle selected for the project

9. The project management plan documents the collection of outputs of the planning processes, which include all the following *except* ______________.

A. How changes will be monitored and controlled

B. How work will be executed to accomplish project objectives

C. The need for and techniques of communication among stakeholders

D. Techniques for negotiating with subcontractors

10. The project manager and the project team must perform multiple actions to execute the project management plan in order to accomplish the work defined in the project scope statement. Which of the following tasks are *not* parts of project execution?

A. Staff, train, and manage team members assigned to the project

B. Develop and prepare a draft copy and get a sign-off on the project charter

C. Document Lessons Learned and implement process improvement activities

D. Perform activities to accomplish project objectives

11. You become aware that without effective change management, a project that you are managing has a limited chance of success. To avoid uncontrolled changes, you set up a change control process, which is primarily concerned with ______________.

A. Maintaining the integrity of the baselines, integrating product and project scope, and coordinating change across knowledge areas

B. Managing project and management contingency reserves

C. Establishing a change control board that oversees the overall project change environment

D. Influencing factors that cause change and determining change that has occurred

12. Most of the project's budget is consumed during ______________.

A. Project plan development

B. Project plan execution

C. Overall change control

D. Project initiation

13. The main purpose of the Develop Project Management Plan process is to ______________.

A. Promote communication among stakeholders

B. Document project assumptions and constraints

C. Define key project objectives

D. Create a document to guide project execution and control

14. In the context of project management, integration includes characteristics of unification, consolidation, articulation, and integrative actions that are crucial to project completion. This is achieved through Project Integration Management processes, which have the main aim of ensuring that the ______________.

A. Project is completed within the approved budget and schedule

B. Proper scope has been defined for the project

C. Various elements of the project are properly coordinated

D. Project will satisfy the needs for which it was undertaken

15. A project management information system (PMIS) is one of the tools and techniques used by the project management team. Which of the following statements is most relevant to the PMIS?

A. It is the software used for schedule development

B. It is the collection of reports generated to support the execution of the project

C. It consists of the tools and techniques used to gather, integrate and disseminate project output

D. It supports all aspects of the project and includes both manual and automated systems

16. All approved changes that can significantly impact the project's scope, cost, schedule, and/or risks must be reflected in the ______________.

A. Performance measurement baselines

B. Project management plan

C. Change management plan

D. Quality assurance plan

17. When developing the project scope statement, project assumptions must be defined and documented as part of the project scope statement. Assumptions generally involve some risk because ______________.

A. Assumptions are based on Lessons Learned

B. Assumptions are based on factors that may not be true, may not be accurate, or may not be available

C. Assumptions are based on constraints

D. Historical information may not be available

18. Which tool or technique is used in the Develop Project Management Plan process?

A. Product analysis

B. Alternatives identification

C. Expert judgment

D. Facilitated workshops

19. In the project management context, integration includes characteristics of unification, consolidation, articulation, and integrative actions that are crucial to the success of the project. Which of the following is the most effective way to reinforce and accelerate the project integration process?

A. Having the sponsor apply direct influence on the project team

B. Through frequent periodic meetings

C. Assigning specific responsibilities to each project employee

D. Using crisis periods to reinforce integration activities

20. Which of the following statements best describes the project's change control board (CCB)?

A. It is required only on large projects.

B. It is headed by the project manager.

C. It is a subcommittee of the Configuration Management Board.

D. It is responsible for approving or rejecting changes to the project baselines.

21. Each project has specific objectives to achieve and deliverables to develop and deliver. In general, project processes fall into two categories: project management processes and ____________ processes.

A. General management

B. Discipline-specific

C. Product-oriented or service-oriented

D. Personnel management

22. Which of the following tasks is *not* a function of configuration management?

A. To identify and document the functional and physical characteristics of an item or system and control any changes to those characteristics

B. To determine if a required change is to be applied to management or project reserves

C. To record and report changes and their implementation status

D. To audit the items and system to verify conformance to requirements

23. Work performance information is an output of which one of these Project Integration Management processes?

A. Direct and Manage Project Execution

B. Perform Integrated Change Control

C. Develop Project Management Plan

D. Close Project or Phase

24. The following items are outputs of the Perform Integrated Change Control process *except* ________.

A. Change request status updates

B. Project management plan updates

C. Project document updates

D. Organizational process assets updates

25. Which statement best describes the Collect Requirements process?

A. Collect Requirements is the process of performing the work defined in the project management plan to achieve the project's objectives.

B. The output of Collect Requirements describes, in detail, the project's deliverables and the work required to create those deliverables.

C. An important part of the Collect Requirements process is ensuring that the schedule is accurately planned to have all stakeholders available at defined times.

D. Collect Requirements is the process of defining and documenting stakeholders' needs to meet the project's objectives.

26. Which of the following actions is part of the project execution activities?

A. Get the sponsor's approval of the project budget

B. Identify the project stakeholders

C. Implement the planned methods and standards

D. Design the project organizational structure

27. Which stage has deliverables as an output?

A. Develop Project Management Plan

B. Monitor and Control Project

C. Perform Integrated Change Control

D. Direct and Manage Project Execution

28. Which of the following statements best describes the impact that requirements have on project scope management?

A. Requirements are used as the basis for the project charter.

B. Requirements are decomposed to create work packages for the project.

C. Requirements become the foundation of the WBS.

D. Requirements are a form of variance analysis for the project.

29. Which one of the following is *not* an output of the Direct and Manage Project Execution process?

A. Change requests

B. Planning processes

C. Work performance information

D. Deliverables

30. Which one of the following is an output of the Direct and Manage Project Execution process?

A. Organizational process asset updates

B. Project management plan

C. Change request status updates

D. Deliverables

31. The following items are part of the information collected routinely to measure and monitor work performance *except* ______________.

A. Costs Incurred

B. Deliverable Status

C. Schedule Forecasts

D. Schedule Progress

32. Which one of the following activities is part of the Monitor and Control Project Work process?

A. Performing activities to accomplish project objectives

B. Tracking and monitoring project risks

C. Staffing management and resource planning

D. Collecting and documenting Lessons Learned

33. Which one of the following activities is *not* part of monitoring and controlling the project work?

A. Adapting approved changes into the project scope, plans, and environment

B. Monitoring implementation of approved changes when and as they occur

C. Comparing actual project performance against the project management plan

D. Providing forecasts to update current cost and current schedule information

34. Change requests could be accepted or rejected or deferred by the relevant change control board, and then updated into the change request log. Change request status updates are an output of which of the following processes?

A. Develop Project Management Plan

B. Develop Project Charter

C. Perform Integrated Change Control

D. Direct and Manage Project Execution

35. Which one of the following activities aims mainly at preserving the integrity of project baselines?

A. Releasing only approved changes

B. Reviewing and approving requested changes

C. Identifying that a change needs to occur or has occurred

D. Reviewing and approving corrective and preventive actions

36. Configuration Management is critical for project success and product integrity. Which one of the following statements reflects the main purpose of configuration status accounting?

A. It provides the basis from which the configuration of products is defined.

B. It establishes that performance and functional requirements have been met.

C. It documents the technical and financial impacts of requested changes.

D. It captures, sorts, and reports the configuration identification list, proposed changes status, and change implementation status.

37. Uncontrolled changes can lead to project failure. The Perform Integrated Change Control process is primarily concerned with managing and controlling changes to the project and the product. All of the following are outputs of Perform Integrated Change Control *except* ______________.

A. Work performance information

B. Change request status updates

C. Project management plan updates

D. Project documents updates

38. As manager of a large project in a large organization under a contract with a customer, you created a multi-tiered structure of a number of Change Control Boards (CCBs). The following statements are all true about the CCBs *except* ____________.

A. A CCB is responsible for reviewing change requests and approving or rejecting those change requests.

B. The roles and responsibilities of the CCB are clearly defined and agreed on by appropriate stakeholders.

C. The CCB may not necessarily record all of the decisions and recommendations that they have made.

D. All CCB decisions are documented and communicated to the stakeholders for information and follow-up actions.

39. What statement is correct about the project's change requests?

A. Change requests are approved by the project's change control board (CCB).

B. Change requests can include corrective action, preventive action, and defect repairs.

C. Approved change requests are implemented when initiating the project.

D. The status of some approved change requests will be updated in the project log.

40. The Close Project or Phase process includes finalizing all activities across all of the project management process groups to formally complete the project or phase. The following activities are necessary for administrative closure *except* ____________.

A. Satisfying completion or exit criteria

B. Transferring the project's products, services, or results to operations

C. Deciding to accept or reject changes to the project or product

D. Collecting project or phase records

41. The procedures used in the Close Project or Phase process include all of the following activities *except* ____________.

A. Reviewing all prior information to ensure that all project work is complete

B. Archiving project information for future use by the organization

C. Investigating and documenting reasons for project termination

D. Verifying and deciding on all submitted change requests

42. Which of the following is *not* considered a part of organizational process assets?

A. Government or industry standards

B. Standardized guidelines and work instructions

C. Organization communication requirements

D. Process measurement databases

43. Closing a project or phase involves using all of the following inputs *except* _______________.

A. Project management plan

B. Accepted deliverables

C. Approved change requests

D. Organizational process assets

44. You are managing a two-year project that spans four different countries. As part of the requirements-gathering process, you have used virtual sessions to compile a list of product requirements. Two differing views remain on one major requirement. You consult with the project sponsor, who reviews the comments and makes a decision favoring one option. She justifies this by stating it will have a greater benefit to most of the end users. What method of decision making is this?

A. Plurality

B. Dictatorship

C. Majority

D. Unanimity

45. Project administrative closure procedures involve all of the following actions and activities *except* _______________.

A. Raising problem reports for nonperforming parts of the system

B. Archiving project information for future use

C. Gathering Lessons Learned

D. Validating that completion and exit criteria have been met

46. You are managing a complex technology project. Because of some unclear requirements in some areas of functionality, the project team has decided to create a rudimentary working model. This model will be used to help the stakeholders review their original requirements and refine them, so that the requirements are sufficient enough to move to a finalized design. This model is an example of which kind of Collect Requirements technique?

A. Focus group

B. Prototyping

C. Observations

D. Change request

47. Under which of the following conditions would the transfer of finished and unfinished deliverables take place?

A. End of a key deliverable

B. End of contract

C. Project termination

D. Legal dispute

48. An activity in the Close Project or Phase process is to collect and archive project or phase files, including all the documentation resulting from the project activities. These include all the following *except* ______________.

A. Project management plan

B. All project team communications

C. Risk and issue register

D. Quality baselines

49. Which one of these items is an input to Define Scope?

A. WBS dictionary

B. Project scope statement

C. WBS

D. Requirements documentation

50. Which one of the following items is an input to Create Work Breakdown Structure (WBS)?

A. Project scope statement

B. Project management plan

C. Project risk register

D. Accepted deliverables

51. Within Project Scope Management, the Collect Requirements process defines and documents stakeholders' needs to meet the project objectives. Tools and techniques used within this process include all of the following *except* ______________.

A. Interviewing stakeholders by talking to them directly, using prepared and/or spontaneous questions

B. Facilitated workshops that bring key cross-functional stakeholders together to define product requirements

C. Development of the initial project management plan baseline in order to manage stakeholders' requirements and expectations

D. Focus groups that use prequalified stakeholders to learn about their expectations concerning the proposed product

52. From the following list, which one is a potential output of the Define Scope process?

A. Requested changes

B. Project document updates

C. Issue log

D. Communications management plan

53. Which one of the following is included in the project scope statement?
   A. Project charter
   B. Project stakeholders
   C. Work breakdown structure
   D. Product scope description

54. What is the lowest level of decomposition in a WBS?
   A. Work package
   B. Component
   C. Deliverable
   D. Work item

55. Which of the following items is contained in the WBS dictionary?
   A. Acceptance criteria
   B. Product description
   C. Code of account identifier
   D. Examples of project components

56. What is the main difference between scope verification and quality control?
   A. Scope verification is concerned with acceptance of the deliverables, whereas quality control is concerned with the correctness of the deliverables.
   B. Quality control is concerned with acceptance of the deliverables, whereas scope verification is concerned with meeting the quality requirements for deliverables.
   C. Scope verification is concerned with testing the required deliverables to ensure that they are feasible, whereas quality control is concerned with testing the final product.
   D. Quality control is performed by groups external to the project, whereas scope verification is performed by the project management team.

57. Inspections can take all of the following forms *except* ______________.
   A. Audits
   B. Reviews
   C. Testing
   D. Walkthroughs

58. The inputs to Control Scope include which of the following?
   A. Requirements traceability matrix
   B. Project budget
   C. Project charter
   D. Project organization

59. Which of the following is a tool or technique used in Control Scope process activities?

A. Product analysis

B. Expert judgment

C. Variance analysis

D. Stakeholder analysis

60. Which of the following items is an output of the Control Scope process?

A. Accepted deliverables

B. Change requests

C. Project scope management plan

D. Approved changes

61. An updated version of which of the following items is an output of the Control Scope process?

A. Change control process

B. Project charter

C. Project scope management plan

D. Project management plan

62. Which one of the following processes is part of the Project Time Management knowledge area?

A. Develop Project Charter

B. Develop Project Management plan

C. Estimate Activity Duration

D. Monitor and Control Project Work

63. Which of the following processes use performance reviews as part of its tools and techniques?

A. Develop Schedule

B. Sequence Activities

C. Control Schedule

D. Define Activities

64. The project manager used the following tools and techniques: decomposition, rolling wave planning, templates, and expert judgment. What Project Time Management outputs will be produced?

A. Project schedule network diagrams and project document updates

B. Activity resource requirements, resource breakdown structure, and project document updates

C. Activity duration estimates and project document updates

D. Activity list, activity attributes, and milestone list

65. Which of the following statements is *not* true about the rolling wave planning technique?

A. It subdivides the project work packages into smaller components called activities.

B. It is a form of progressive elaboration in which future work is planned at a higher level of the WBS.

C. It decomposes work packages to a milestone level when information is less defined.

D. It decomposes work packages into activities as more becomes known about upcoming activities.

66. What is the shortest duration of the following project, according to the precedence diagramming method (PDM)?

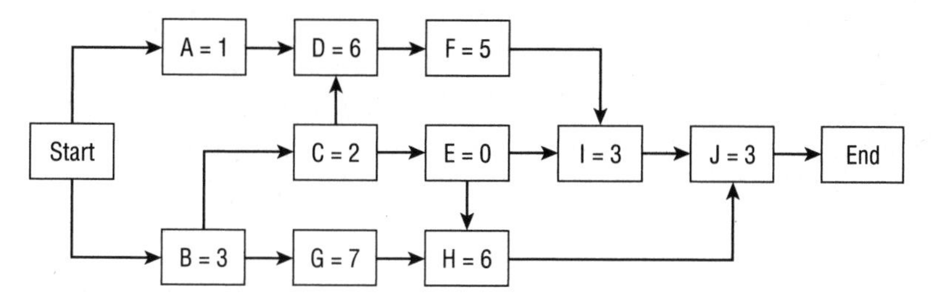

A. 18 units

B. 22 units

C. 19 units

D. 11 units

67. Referring to the following network diagram, what is the critical path?

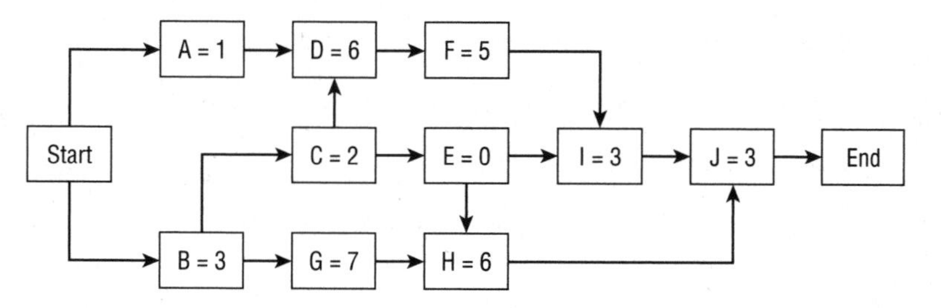

A. ADFIJ

B. BCEIJ

C. BCDFIJ

D. BGHJ

68. You have been assigned to manage a complex construction project. No one on your project team has deep knowledge or experience in the construction industry. You have hired a local consultant to deliver an awareness seminar for your project team to familiarize them with the local building codes and specialized construction techniques. Which process will this knowledge help you and your team to complete?

A. Estimate Activity Durations

B. Define Activities

C. Sequence Activities

D. Estimate Activity Resources

69. Which of the following terms about the project equipment and material resources is an input to the Estimate Activity Duration process?

A. Price

B. Availability

C. Make

D. Supplier

70. In activity duration estimating, three-point estimates are based on determining three types of estimates. Which one of the following is *not* one of the types of estimates used in the three-point estimating technique?

A. Most likely

B. Least likely

C. Optimistic

D. Pessimistic

71. The critical chain method is a schedule network analysis technique that modifies the project schedule by adding duration buffers that are non-work schedules. The purpose of these buffers is to ______________.

A. Inflate the project effort, schedule, and plan

B. Balance the project resources relative to the project milestones

C. Maintain focus on the planned activity durations

D. Help manage the total float of network paths

72. An assistant to the senior project manager produced an output that has specific baseline start and end dates that will be included in the project management plan. This particular output, reviewed and approved by the project management team, was developed from the schedule network analysis. What is the name of this Project Time Management output?

A. Schedule baseline

B. Activity list

C. Project schedule network diagrams

D. Activity duration estimates

73. From the following list, which is an input to the Control Schedule process?

A. Work performance measurements

B. Change requests

C. Work performance information

D. Project management plan updates

74. From the following list, which one is an item on the resource calendar?

A. Weekends

B. Resource skills

C. Vacation periods

D. Contract timelines

75. In which of the following situations is root cause analysis used?

A. Closing the project

B. Estimating the project

C. Following an emergency fix

D. Assessing a change request

76. In your project, you share a skilled resource with other projects, and you must review your project schedule to reflect this resource sharing. You plan to do this through resource leveling. Resource leveling will often affect the project by making the schedule _______________.

A. Shorter

B. Longer

C. More responsive to customer needs

D. Less flexible

77. In your project plan, you budgeted three weeks of effort for consultants, but the job was done in only two weeks. Calculate the percentage of variance.

A. 150%

B. 33%

C. 75%

D. 67%

78. On November 1, $1,000 worth of work on Task A was planned to be complete. The work performed at the planned rate was valued only at $850. Calculate the schedule variance.

A. –$100

B. $100

C. –$150

D. $150

79. Which of the following costs are *not* relevant during cost budgeting decision making?

A. Direct labor costs

B. Overhead costs

C. Material costs

D. Sunk costs

80. You are managing a project with a budget of $5 million. You are two months into the project, and the cost figures are as shown in the following table:

Period	Actual Cost	Planned Value	Earned Value
Month 1	1,250,000	1,100,000	1,000,000
Month 2	500,000	600,000	750,000
Month 3	–	2,500,000	–
Month 4	–	800,000	–

What statement is *not* true about the project at the end of Month 1?

A. The project is behind schedule.

B. The project is over budget.

C. The cost performance index is 1.25.

D. The schedule variance is –$100,000.

81. You are managing a project with a budget of $5 million. You are two months into the project, and the cost figures are as shown in the following table:

Period	Actual Cost	Planned Value	Earned Value
Month 1	1,250,000	1,100,000	1,000,000
Month 2	500,000	600,000	750,000
Month 3	–	2,500,000	–
Month 4	–	800,000	–

What statement is true about the project?

A. At the end of Month 2, the project is on budget.

B. At the end of Month 2, the project is behind schedule.

C. For Month 2, the cost performance index is .66.

D. For Month 2, the schedule variance is –$300,000.

82. A project consists of the three activities shown in the following table. Assuming that future work will be performed according to the budget, the estimate at completion (EAC) at this point in the project is ________.

Activity	% Complete	Planned Value	Earned Value	Actual Cost
A	100%	1,000	1,000	1,200
B	50%	1,000	500	700
C	0%	1,000	–	–

A. \$3,800

B. \$3,700

C. \$3,200

D. \$3,000

83. You were asked to prepare an estimate at completion (EAC) based on the work performed at the present cost performance index (CPI). Which EAC formula should you use?

A. EAC = AC + Bottom-Up ETC

B. EAC = BAC / Cumulative CPI

C. EAC = AC + BAC – EV

D. EAC = AC + [(BAC – EV) / Cumulative CPI × Cumulative SPI)]

84. Which of the following estimating techniques has been shown to be the least accurate?

A. Analogous

B. Parametric

C. Three-point

D. Bottom-up

85. Given the following graphic report on a project, what can be said about the project performance at the point in time indicated as Today's Date?

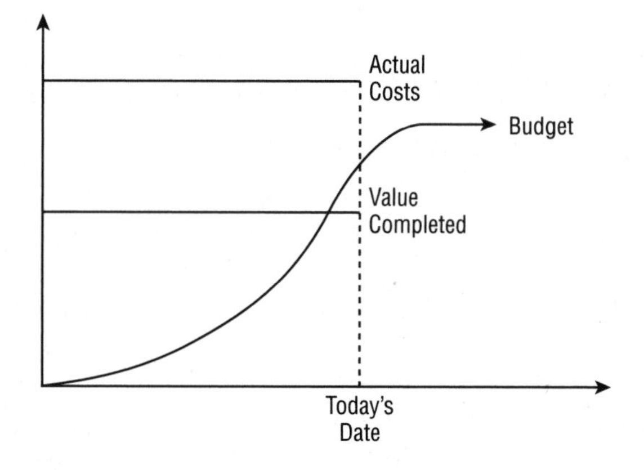

A. The project is over budget but ahead of schedule.

B. The project is over budget and behind schedule.

C. The project is under budget and ahead of schedule

D. Additional information is needed before conclusions can be made about budget and schedule performance.

86. What is the difference between management reserves and contingency reserves?

A. There is no difference; they mean the same thing.

B. Management reserves are applied to schedule and contingency reserves are applied to cost.

C. Contingency reserves are included in the project's cost performance baseline and management reserves are not.

D. Management reserves cover known unknowns, whereas contingency reserves cover unknown unknowns.

87. Which of the following statements about quality and grade is correct?

A. High quality is always a problem and low grade is not a problem.

B. Low quality is always a problem and low grade may not be.

C. Both low quality and low grade are always a problem.

D. Low quality and high grade are always a problem.

88. The following are all direct advantages of continuous process improvement *except* ______________.

A. It reduces waste

B. It reduces non-value-added activities

C. It increases process efficiency

D. It eliminates project risks

89. Which of the following is *not* part of the quality costs?

A. Costs of investment made to prevent nonconformance to requirements

B. Costs incurred by appraising the product or service

C. Costs of productivity bonuses to the project team

D. Costs of services to ensure compliance with mandatory standards

90. Which of the following descriptions is correct for the Perform Quality Assurance process?

A. The process of auditing the quality requirements and the results from quality control measurements to ensure appropriate quality standards are used

B. The application of planned systematic user acceptance testing to ensure customer acceptance of the project using various tools and techniques

C. The application of planned systematic verification techniques to ensure the correctness of all the project's work products to provide final delivery of deliverables

D. The planning and implementing of inspections to ensure the correctness and functionality of all the delivered modules in accordance with approved specifications

91. Modern quality management complements project management in that both disciplines recognize the importance of all of the following *except* _______________.

A. Prevention over inspection

B. Management responsibility

C. Procurement management

D. Continuous improvement

92. The primary components of the quality management function are _______________.

A. Quality planning, quality control, and quality assurance

B. Qualitative and quantitative measurement

C. Qualitative assessment of the product or service

D. Acceptance sampling and statistical process control

93. Project quality assurance _______________.

A. Includes policing the conformance of the project team to specifications

B. Is a managerial process that defines the organization, design, resources, and objectives of quality management

C. Provides confidence that the project will satisfy the quality standards that have been defined for the project

D. Provides the project team and stakeholders with standards by which project performance is measured

94. The _______________ has ultimate responsibility for quality control.

A. Project engineer

B. Project manager

C. Functional manager

D. Quality manager

95. Project quality audits confirm the implementation of the following *except* _______________.

A. Approved change requests and defect repairs

B. Management reserve

C. Corrective actions

D. Preventive actions

96. Which one of the following is an output of performing quality assurance?

A. Rejected change requests

B. Quality cost report

C. Testing error report

D. Change requests

97. Seven of the tools and techniques of quality control are known as the Seven Basic Tools of Quality. Which one uses rank ordering to focus corrective action?

A. Pareto chart

B. Histogram

C. Run chart

D. Scatter diagram

98. Which tool of the Seven Basic Tools of Quality is used to show trends of a process over time, variation over time, or decline or improvement of a process over time?

A. Scatter diagram

B. Histogram

C. Run chart

D. Pareto chart

99. Which one of the following processes is *not* part of the project's human resource–related processes?

A. Acquire Project Team

B. Develop Human Resource Plan

C. Administer Payroll

D. Manage Project Team

100. The RACI chart is an example of a Responsibility Assignment Matrix (RAM). Which of the following is *not* a correct description for describing roles and expectations?

A. Responsible

B. Assign

C. Consult

D. Inform

101. Project Human Resource Management includes which of the following processes?

A. Develop Human Resource Plan, Acquire Project Team, Develop Project Team, and Manage Project Team

B. Hire Human Resources, Train Staff, Mentor Project Team, and Recognize Success

C. Recruit Personnel, Maintain Labor Relations, and Administer Payroll

D. Build the Team, Communicate Roles, and Distribute Labor

102. Which one of the following is *not* an input of human resource planning?

A. Activity resource requirements

B. Work breakdown structure (WBS)

C. Enterprise environmental factors

D. Organizational process assets

103. Which of the following statements is *not* a factor relevant to the project's human resource planning?

A. Which departments will be involved in the project

B. What the team members will cost

C. What distances separate the people that will be involved in the project

D. The skills and experience of the team members

104. The human resource plan component (which can be formal or informal) that describes when and how the project human resource requirements will be met is called the ______________.

A. Organizational breakdown structure

B. Responsibility assignment matrix

C. Staffing management plan

D. Resource assignment chart

105. Project team members are drawn from all available sources, both internal and external. The following characteristics influence or direct the assignments of the project team members *except* ______________.

A. Availability

B. Cost

C. Experience

D. Hobbies

106. All of the following tools and techniques are used in team development *except* ______________.

A. Team-building activities

B. Recognition and rewards

C. Stakeholder analysis

D. Training

107. Which one of the following behaviors is *not* a behavior indicating effective teamwork?

A. Team members share information and resources

B. Team members socialize together after work hours

C. Team members assist one another when workloads are unbalanced

D. Team members communicate in ways that fit individual preferences

108. Unplanned training takes place as a result of all of the following *except* ________________.

A. Observation

B. Conversation

C. Project performance appraisal

D. Human resource plan

109. Which of the following statements is *not* true about the Develop Project Team process?

A. Team development improves the trust between individual team members.

B. Team development occurs throughout the project life cycle.

C. All project team members share responsibility for enforcing team ground rules.

D. A team member's recognition and rewards system is a recommended approach for team development.

110. Which of the following statements is *not* a characteristic of virtual teams?

A. Team communications tend to be dominated by email and other electronic forms.

B. Teams tend to be groups of people with a shared goal who spend little time meeting face to face.

C. Team members are concerned with managing diverse stakeholder views and expectations.

D. It is possible to include people with mobility handicaps.

111. Interpersonal conflicts can be difficult to resolve in a project environment because ________________.

A. The project manager does not have authority to resolve these conflicts

B. The limited time of the project does not allow for resolution of these types of conflicts

C. These conflicts often involve ethics, morals, and value systems

D. The organizational structure of the project hinders resolution

112. The indicators of team effectiveness include all of the following statements *except* ________________.

A. Increased conflict between the project manager and the project personnel

B. Evident improvements in skills

C. Continuous improvements in competencies

D. Reduced staff turnover

113. Which of the following interpersonal skills are *not* required of project managers?

A. Understanding compensation plans, benefits, and career paths

B. Developing a vision of the future and the necessary strategies to achieve it

C. Aligning people through focused communication

D. Helping people to energize themselves to overcome barriers to change

114. Which of the following factors contributes the most to effective team communication?

A. Performance appraisals

B. External feedback

C. Conflict resolution

D. Co-location

115. The project management team performs the following activities *except* _______________.

A. Managing conflicts affecting project performance

B. Resolving problems affecting project performance

C. Establishing salary levels of the team members

D. Appraising performance of team members

116. You are now in the stage of monitoring and controlling a project. Which one of the following tasks is *not* part of the Manage Project Team process?

A. Tracking team performance

B. Providing feedback

C. Resolving issues

D. Allocating pay raises

117. The purpose of developing the project team is to increase the effectiveness of the team through all of the following *except* _______________.

A. Improving the skills of the team members through exchanging knowledge and experience

B. Encouraging team members to go to lunch together and have fun

C. Encouraging team members to share information and resources

D. Improving the feelings of trust and cohesion among team members

118. Which of the following beliefs about conflict is typically held by projectized organizations?

A. Conflict is inevitable.

B. Conflict is bad.

C. Conflict is determined by the structure of the system.

D. Conflict is caused by the relationships among components.

119. Which of the following statements is true about the nature of conflicts?

A. The overall level of conflict remains relatively constant over the project life cycle.

B. Schedules and priorities are the most common sources of project conflicts.

C. Interpersonal situations are the most common sources of project conflicts.

D. Project managers typically favor the compromise approach to handling conflicts.

120. The 360-degree feedback principle in project performance appraisal implies all of the following *except* ______________.

A. Feedback from the project manager on the performance of the project team member

B. Feedback from the project team member's personal friends and family on his or her personality

C. Feedback from the project team member's superiors, managers, and supervisors

D. Feedback from the project team member's superiors, peers, and subordinates

121. The major processes of Project Communications Management are ______________.

A. Identify Stakeholders, Plan Communications, Distribute Information, Manage Stakeholder Expectations, Report Performance

B. Identify Stakeholders, Plan Communications, Response Planning, Progress Reporting and Information

C. Identify Stakeholders, Plan Communications, Distribute Information, Schedule Reporting, and Stakeholder Analysis

D. Identify Stakeholders, Plan Communications, and Change Reporting, Project Records, and Acceptance

122. Which one of the following Project Communications Management processes is part of the Monitoring and Controlling process group?

A. Plan Communications

B. Manage Stakeholder Expectations

C. Report Performance

D. Distribute Information

123. In the project's communications management, which one of the following is *not* a key component of the basic model of communication?

A. Noise

B. Message

C. Encoding

D. Messenger

124. Which one of the following factors does *not* affect a communications management plan?

A. Geographical distribution of project staff

B. Project length

C. Executive requirements

D. Available technologies

125. Your client, with whom you have good relations, asks you verbally to provide him with office supplies. This request is not supported under your contract. Which medium would be best to use to initially respond to your client's request?

A. Presentation

B. Telephone call

C. Oral request

D. Formal letter

126. You were part of a team that worked on a complex systems integration project. The technical architect left the company during the project. Six months later, the technical architect contacts you and asks for an electronic copy of the project technical solution document because he has only a paper copy and is working on a similar project. You should ______________.

A. Send him the details because he developed the original infrastructure and therefore knows all the information already

B. Not send him the details but invite him to dinner so he can review it with you

C. Not send him the files; he does not have a legitimate need to know

D. Send him the files along with a confidentiality agreement that he must sign

127. Which one of the following statements is *not* part of the communications management plan?

A. To whom information will be distributed

B. What, how, and how often information will be gathered

C. What methods will be used to distribute information

D. All memos, correspondence, and reports from project personnel

128. Which of the following techniques is used in the Distribute Information process?

A. Stakeholder buy-in

B. Executive interviews

C. Communication methods

D. Channel evaluation

129. The project sponsor requests a monthly meeting with the project manager to allow the project manager to present status information in person, rather than sending a report by email. This additional activity requires an update to ______________.

A. The project charter

B. The communications management plan

C. The stakeholder management plan

D. The project's change control process

130. There are different methods of communications: oral and written, formal and informal. Choice of the communication method depends on the situation. Formal written correspondence is mandated in which one of the following situations?

A. The product undergoes casual in-house testing.

B. The client requests additional work not covered under the contract.

C. The project manager calls a meeting.

D. A customer executive requests a review of the project.

131. Performance reports can include all of the following *except* ______________.

A. Results of the ISO audit

B. Results of cost control

C. Results of quality control

D. Results of schedule control

132. The outputs of the Report Performance process include all of the following *except* ______________.

A. Organizational process assets updates

B. Performance reports

C. Change requests

D. Product analysis

133. In which project management process group is the Identify Stakeholders process performed?

A. Initiating

B. Planning

C. Executing

D. Monitoring and Controlling

134. A formal written response is most suitable when ____________.

A. Discussing work assignments

B. Directing the contractor to make changes

C. The interpreter has a short memory

D. Contacting the client is necessary

135. Which one of the following tools is *not* one of the tools and techniques used in performance reporting?

A. Reporting systems

B. Variance analysis

C. Project planning sessions

D. Status review meetings

136. Which of the following statements is *not* true about actively managing stakeholders in a project?

A. It will increase the likelihood that the project will stay on track.

B. Project team members will spend unnecessary efforts in communicating with key stakeholders.

C. The project manager is responsible for stakeholder management.

D. It helps limit disruption during the project by proactively managing issues.

137. As the project manager, you want to impose a complex accounting method by which subordinates will calculate project costs. You have time for only one approach. Which type of communication would be most effective in communicating this method?

A. Verbal, face-to-face

B. Verbal, telephone

C. Written

D. Nonverbal

138. It is important for a project manager to assess the complexity of the communication requirements of the project by calculating the number of communication channels. If the project has eight stakeholders, what is the total of number of potential communication channels?

A. 16

B. 28

C. 64

D. 256

139. The project you are managing has a variety of stakeholders, including a number of contractors. The project's issue log is an input to ______________.

- **A.** Team building
- **B.** Risk management
- **C.** Stakeholder management
- **D.** Schedule management

140. As stakeholder requirements are identified, issues will be logged for resolution. Once resolved, they will be documented with a number of information items. The following examples are possible information items recorded in the issue log *except* ______________.

- **A.** Follow-on action agreed on by the relevant stakeholders
- **B.** Root cause analysis of the issue
- **C.** Results of negotiations with functional managers
- **D.** Results of the project risk analysis

141. Which one of the following is the best way to prepare a response to unknown risks?

- **A.** Ignore them
- **B.** Prepare a risk response
- **C.** Plan to treat them the same way as the known risks
- **D.** Allocate management reserves against them

142. Which of the following statements is *false* about risk management plans?

- **A.** Risk management plan is a subset of the project management plan.
- **B.** The Plan Risk Management process produces the risk management plan as an output.
- **C.** The risk management plan should be updated frequently.
- **D.** Risk management plans may include budgeting, timing, and tracking details.

143. Which one of the following techniques is used in the Identify Risks process?

- **A.** WBS
- **B.** Delphi
- **C.** Decomposition
- **D.** Structured analysis

144. A risk register typically contains the following *except* ______________.

- **A.** List of potential responses
- **B.** Result of root cause analysis
- **C.** Work breakdown structure
- **D.** List of identified risks

145. Which of the following risk categorization grouping criteria leads to the most effective risk responses?

A. Root causes

B. Source of risk

C. Areas of WBS

D. Project phase

146. In quantitative risk analysis and modeling, sensitivity analysis will typically use a(n) _______________ to compare the relative importance and impact of variables with a high degree of uncertainty against variables that are more stable.

A. Decision tree diagram

B. Tornado diagram

C. Expected monetary value

D. Control chart

147. In any project, risks can be negative (threats) or positive (opportunities). Which of the following is *not* a strategy for dealing with negative risks?

A. Exploit

B. Mitigate

C. Avoid

D. Transfer

148. In any project, risks can be negative (threats) or positive (opportunities). Which of the following is *not* a strategy for dealing with positive risks?

A. Exploit

B. Enhance

C. Share

D. Increase

149. Which one of the following is an input to the Monitor and Control Risks process?

A. Updated WBS

B. Project management plan

C. Change requests

D. Project organization chart

150. Which one of the following is *not* an input, tool, or technique used in risk monitoring and control?

A. Status meetings

B. Project staff rewards scheme

C. Variance and trend analysis

D. Technical performance measurement

151. Project Procurement Management deals with purchasing or acquiring products, services, or results from external providers to enable the project team to perform the project work. All of the following are procurement processes *except* ______________.

A. Plan Procurements

B. Conduct Procurements

C. Estimate Costs

D. Close Procurements

152. Which of the following is *not* typically a content of the procurement management plan?

A. Types of contract to be used

B. Standardized procurement document

C. Constraints and assumptions

D. The selected sellers

153. Which of the following contract types has the lowest risk for the buyer?

A. Cost-plus-fixed-fee

B. Fixed-price

C. Cost-plus-incentive-fee

D. Time and materials

154. Which one of the following is *not* a cost-reimbursable type of contract?

A. Cost-plus-fixed-fee

B. Fixed-price

C. Cost-plus

D. Cost-plus-percentage-fee

155. Which one of the following usually is *not* a candidate for sellers' evaluation criteria during the Plan Procurements process?

A. Sellers' management approach

B. Sellers' technical approach

C. Sellers' financial capacity

D. Sellers' stock price

156. Which of the following is *not* an output of the Conduct Procurements process?

A. Selected sellers

B. Contract award

C. The requested product

D. Project management plan updates

157. Which one of the following is *not* a tool or technique used in the Conduct Procurements process?

A. Independent estimates

B. Bidder conferences

C. Procurement negotiation

D. Claims administration

158. The procurement contract award is an output of which of the following Project Procurement Management processes?

A. Administer Procurements

B. Close Procurements

C. Plan Procurements

D. Conduct Procurements

159. Which one of the following is *not* an output of the Conduct Procurements process?

A. Procurement statement of work

B. Project document updates

C. Resource calendars

D. Change requests

160. Which one of the following is *not* part of the tools and techniques used in the Administer Procurements process?

A. Contract change control system

B. Inspection and audits

C. Records management system

D. Proposal evaluation techniques

161. Which one of the following items is *not* subject for contract discussion during procurement negotiations?

A. Responsibilities and authorities

B. Overall schedule, payment, and price

C. Technical solutions

D. Project personnel

162. The records management system, used by the project manager to manage contract documentation and records, consolidates all of the following *except* ______________.

A. Set of processes

B. Automation tools

C. Payment system

D. Related control functions

163. You are managing a project that will involve third-party sellers. The procurement documents have been released, and several questions and requests for clarifications have been submitted. To be fair, you and the procurement officer will hold a real-time meeting with all prospective sellers to answer their questions. This is known as ______________.

A. Proposal evaluation techniques

B. A bidder conference

C. Procurement negotiations

D. Inspections and audits

164. Your company has signed a contract with a third party for your project. Late in the project, you and the third party believe that a contract change is necessary. Which of the following statements is true?

A. Contract change is not possible at this late stage of the project.

B. You should refer to a neutral organization.

C. Change is possible by following the change control process.

D. Trying to dissuade the third party from any changes is an appropriate action at this point.

165. Which of the following is *not* an element of procurement management?

A. Plan Project

B. Plan Procurement

C. Conduct Procurement

D. Administer Procurement

166. Which of the following statements is *not* true about contracts with selected contractors?

A. The contract is subject to remedy in the courts in case of legal disputes.

B. The contract does not include any specific roles and responsibilities.

C. The contract is a mutually binding legal agreement between the buyer and the seller.

D. The contract obligates the buyer to compensate the seller, and the seller is obligated to provide specified products or services.

167. Contracts are legally binding documents between the buyer and the seller. Complete this sentence with one of the following clauses: Many organizations perform contract administration ______________.

A. Separate from the project

B. As part of the project

C. As part of their personnel department

D. External to the organization

168. As part of your project, the team is investigating which, if any, work packages will be performed by third-party sellers and which will be performed internally by the project team, taking into account resource availabilities and skills. This is known as (a) _______________.

A. Procurement statement of work

B. Requirements documentation

C. Procurement audit

D. Make-or-buy decision

169. Of these options, which one is *not* an input to the Conduct Procurements process?

A. Organizational process assets

B. Project management plan

C. Selected sellers

D. Qualified sellers list

170. Which of the following project management processes is *not* among those that directly interact with the Administer Procurements process?

A. Direct and Manage Project Execution

B. Initiate the Project

C. Perform Quality Control

D. Perform Integrated Change Control

171. Of these options, which one is *not* an input to the Administer Procurements process?

A. Work performance information

B. Performance reports

C. Inspections and audits

D. Approved change requests

172. Procurement performance reviews are one of the tools and techniques used to support the Administer Procurements process. They are used by buyers for several purposes, including all of the following *except* _______________.

A. To identify performance successes or failures

B. To verify progress in accordance with the procurement statement of work

C. To quantify the seller's demonstrated ability to perform work

D. To select the appropriate sellers for the project

173. The Close Procurements process involves all of the following activities *except* _______________.

A. Verification of all work and deliverables

B. Impact assessment of requested changes

C. Updating records to reflect the final results

D. Archiving the project records

174. Early termination of a contract is a special case of procurement closure. Which one of the following statements is *not* true about early termination of a contract?

A. Early termination of a contract may take place only through legal action.

B. Early termination of a contract may result from mutual agreement between the buyer and seller.

C. Rights and responsibilities in the event of early termination are contained in the contract.

D. In some cases, the buyer may have to compensate the seller for any accepted work.

175. Organizational process assets updates resulting from the Close Procurements process could cover all of the following *except* _______________.

A. Archive of the procurement file

B. Final acceptance of deliverables

C. Gathering of Lessons Learned

D. Changes to the contract

176. Contested changes are those requested changes for which the buyer and seller cannot agree on compensation for the change. Which one of the following terms is *not* a term used to refer to contested changes?

A. Claims

B. Disputes

C. Refund requests

D. Appeals

177. Which document includes the procurement contract, seller-developed technical documentation, seller performance reports, and results of contract-related inspections?

A. Procurement management plan

B. Procurement contract award

C. Closed procurements

D. Procurement documentation

178. Which one of the following is *not* a tool or technique of the Close Procurements process?

A. Negotiated settlements

B. Records management system

C. Procurement audits

D. Expert judgment

179. Which one of the following is *not* an output of the Close Procurements process?

A. Project management plan

B. Closed procurements

C. Procurement file

D. Lessons learned

180. The procurement rules and requirements for the formal deliverables acceptance are usually defined in which one of the following documents?

A. The change control system

B. A statement of work

C. The testing strategy

D. The contract

181. The requirements for contract closure usually are included in which of the following documents?

A. Configuration management plan

B. Requirements document

C. Project plan

D. Procurement management plan

182. How does the buyer inform the seller that the deliverables have been accepted or rejected?

A. Oral confirmation

B. Written communication

C. Formal written notice

D. Change request

183. Professional project managers are expected to align their daily performance with four values that have been identified as the most important to the project management community. Which of the following terms is *not* part of those four values?

A. Responsibility

B. Hard work

C. Respect

D. Honesty

184. In the middle of a large construction project, a questionable behavior was observed with one of the project team members. Although the project sponsor is aware that a *Code of Ethics and Professional Conduct* applies to all PMI® members as well as some non-PMI® members, she is not sure which of the following nonmembers it does *not* apply to. What should you tell her?

A. Nonmembers who hold PMI® certificates

B. Nonmembers who serve PMI® in a volunteer capacity

C. Nonmembers who currently manage projects

D. Nonmembers who aspire to become PMI® certified

185. The purpose of *PMI® Code of Ethics and Professional Conduct* is that practitioners of project management should do all the following *except* ______________.

A. Be committed to doing what is right and honorable

B. Set high standards for ourselves

C. Apply our standards in a flexible way so that we do not lose our jobs

D. Aspire to meet these standards in all aspects of our lives as well as at work

186. The values that formed the basis of global project management community decision making and guided decision makers' actions include all of the following *except* ______________.

A. Focus group discussions

B. A PMI® direct email survey of 10,000 project managers and projects

C. Two Internet surveys involving practitioners, members, volunteers, and people holding PMI® certification

D. Management consultants hired by the PMI® to develop these values

187. Which of the following statements is *not* true regarding mandatory versus aspirational standards in the *PMI® Code of Ethics and Professional Conduct*?

A. Aspirational standards establish firm requirements.

B. Practitioners who do not conduct themselves in accordance with mandatory standards will be subject to disciplinary procedures.

C. Aspirational standards establish conduct that we should strive to uphold.

D. Adherence to aspirational standards is not easily measured.

188. You notice that a European team leader on one of your projects has been communicating with an Asian member of the team in an abusive manner. Your duty as a project manager is to ______________.

A. Ignore this behavior to keep the project going smoothly

B. Ask the team leader to stop this behavior

C. Ignore this behavior to keep your good relationship with the team intact

D. Record the facts regarding this to substantiate the team leader's behavior and report the team leader to his or her functional manager

189. As a project manager, in which of the following situations would you pursue disciplinary action?

A. A member of the project team reports unethical behavior of another team member

B. A member of the project team complains that he or she is the subject of unethical abuse by another member of the team

C. A member of the project team retaliates against a colleague who raised an ethical complaint against him or her

D. A member of the project team files an ethics-related complaint without substantiating it with facts

190. You have been asked by the program manager to manage a highly sensitive project related to building a nuclear reactor. The project requires specialized knowledge about the nuclear industry. You have never worked in that industry before, although you managed a number of defense projects in other conventional areas. Which of the following is the most appropriate action for you to consider?

A. Take on the project and start self-learning about the nuclear industry without letting anyone know

B. Inform the program manager that you do not have experience in working in the nuclear industry and that you are prepared to learn on the job

C. Tell your manager that you will learn on the job and ask him or her not to inform the customer that you do not have the right knowledge, so your company will not lose the opportunity

D. Refuse the project because you do not have the right experience

191. You completed an advanced technology project for one customer. You moved on to manage a related project for another customer, and you wanted to copy some of the work products to use in the second project without telling the first customer. What should you do?

A. Copy the work products and ask your management not to tell the first customer

B. Copy the work products without telling your management or the customer

C. Do not copy the work products before asking the first customer for permission

D. Ask a member of the first project team to copy the work products for you without informing anyone else

192. To which one of the following areas in the *PMI® Code of Ethics and Professional Conduct* does active listening belong?

A. Respect/mandatory

B. Responsibility/mandatory

C. Responsibility/aspirational

D. Respect/aspirational

193. Abusive behavior violates which one of the *PMI® Code of Ethics and Professional Conduct* values?

A. Respect/mandatory

B. Fairness/mandatory

C. Respect/aspirational

D. Fairness/aspirational

194. You are managing a large international project that pays well. Your uncle asked you to hire his son in the project, even though your cousin does not have the necessary skills. Hiring your cousin would constitute which type of *PMI® Code of Ethics and Professional Conduct* conflict of interest?

A. Fairness/mandatory

B. Responsibility/aspirational

C. Honesty/mandatory

D. Respect/mandatory

195. One of the value standards in the *PMI® Code of Ethics and Professional Conduct* requires practitioners of global project management to "report unethical or illegal conduct to appropriate project management and if necessary to those affected by the conduct." To which of the following values of the *PMI® Code of Ethics and Professional Conduct* does this standard belong?

A. Honesty/mandatory

B. Respect/aspirational

C. Fairness/aspirational

D. Responsibility/mandatory

196. Several accidents in a nuclear energy project led to the loss of life on a number of occasions, under the same project managers. Which of the following *PMI® Code of Ethics and Professional Conduct* values did the project managers betray?

A. Fairness

B. Responsibility

C. Respect

D. Honesty

197. A newly assigned project manager discovered a discrepancy in the budget that was prepared by the previous project manager. If he communicates the discrepancy to the project sponsor, the project will most likely be canceled immediately and his contract will be terminated prematurely. He just bought a new house and his wife is expecting triplets.

Which value would he be violating the *most* if he delays the disclosure of the budget discrepancy until he can find a replacement contract?

A. Responsibility

B. Respect

C. Fairness

D. Honesty

198. As a professional project manager, you were faced with a decision involving potential conflicts of interest. You believe that your decision was correct and advantageous for the project. All of the following statements are necessary preconditions for engaging in this decision-making process *except* _______________.

A. Making full disclosure to the affected stakeholders

B. Having an approved mitigation plan

C. Obtaining the consent of the stakeholders to proceed

D. Reporting the situation to the PMI®

199. A project manager promised project team members bonuses that he never intended to deliver. Which type of value does this behavior represent?

A. Responsibility/aspirational

B. Responsibility/mandatory

C. Fairness/aspirational

D. Honesty/aspirational

200. As manager of a large project, you needed to recruit a junior project manager with specific skills. After the new junior project manager had a few months on the job, you realized that he did not have all the skills he claimed to have. Which of the following *PMI® Code of Ethics and Professional Conduct* value standards did his behavior violate?

A. Fairness/mandatory

B. Responsibility/mandatory

C. Respect/aspirational

D. Responsibility/aspirational

201. Thomas is managing an international project that will span many years. Although the raw materials are sourced from one country, he is anticipating that inflation will impact the labor costs in at least two other countries where the project will be implemented. The price of fuel, a key component in the project, may fluctuate as well. What would be the *most* appropriate contract given the nature of his project?

A. Cost-reimbursable contracts

B. Firm Fixed-Price (FFP)

C. Fixed-Price Incentive Fee (FPIF)

D. Fixed-Price with Economic Price Adjustment (FP-EPA)

202. If the procurement decision will be based primarily on price, which procurement document would be *most* appropriate?

A. Invitation for bid (IFB) or request for quotation (RFQ)

B. Request for proposal (RFP) or request for information (RFI)

C. Request for information (RFI) or invitation for bid (IFB)

D. Invitation for bid (IFB) or request for proposal (RFP)

203. Which of the following is *not* an input of the Direct and Manage Project Execution process?

A. Project charter

B. Project management plan

C. Approved change requests

D. Enterprise environmental factors

204. Preassignment, negotiation, acquisition, and virtual teams are tools and techniques of the ______________ process.

A. Build Project Team

B. Acquire Project Team

C. Develop Team Structure

D. Procure Resources

205. The contract life cycle consists of which four stages?

A. Forming, norming, storming, and adjourning

B. Preassignment, negotiation, acquisition, and payment

C. Requirement, requisition, solicitation, and award

D. Plan, assure, control, and close

Chapter 8

Answers to Practice Test C

Answer Key for Practice Test C

1. C
2. C
3. B
4. D
5. C
6. A
7. A
8. C
9. D
10. B
11. A
12. B
13. D
14. C
15. D
16. B
17. B
18. C
19. D
20. D
21. C
22. B
23. A
24. D
25. D
26. C
27. D
28. C
29. B
30. D
31. C
32. B
33. A
34. C
35. A
36. D
37. A
38. C
39. A
40. C
41. D
42. A
43. C
44. B
45. A
46. B
47. C
48. B
49. D
50. A
51. C
52. B
53. D
54. A
55. C
56. A
57. C
58. A
59. C
60. B
61. D
62. C
63. C
64. D
65. A
66. B
67. C
68. D
69. B
70. B
71. C
72. A
73. C
74. B
75. C
76. B
77. B
78. C
79. D
80. C
81. A
82. A
83. B
84. A
85. B
86. C
87. B
88. D
89. C
90. A
91. C
92. A
93. C
94. B
95. B
96. D
97. A
98. C
99. C
100. B
101. A
102. B
103. A
104. C
105. D
106. C
107. B
108. D
109. A
110. C
111. C
112. A
113. A
114. D
115. C
116. D
117. B
118. A
119. B
120. B
121. A
122. C
123. D
124. B
125. D
126. C
127. D
128. C
129. B
130. B
131. A
132. D
133. A
134. B
135. C
136. B
137. A
138. B
139. C
140. D
141. D
142. C
143. B
144. C
145. A
146. B
147. A
148. D
149. B
150. B
151. C
152. D
153. B
154. B
155. D
156. C
157. D
158. D
159. A
160. D
161. D
162. C
163. B
164. C
165. A
166. B
167. A
168. D
169. C
170. B
171. C
172. D
173. B
174. A
175. D
176. C
177. D
178. D
179. A
180. D
181. D
182. C
183. B
184. C
185. C
186. B
187. A
188. D
189. C
190. B
191. C
192. D
193. A
194. A
195. D
196. C
197. D
198. D
199. A
200. B
201. D
202. A
203. A
204. B
205. C

1. C. Managing a project always involves managing multiple processes across multiple process groups. Without effective integration of the project processes across different groups, the project is likely to fail. The project may be affected by government legislation, but you do not need to integrate with government plans. Project integration management is about coordinating the various processes within the project management process groups.

   Reference: *PMBOK® Guide,* 4th Ed., page 71

2. C. Perform Integrated Change Control, Develop Project Charter, and Close Project or Phase are all processes that belong to the Project Integration Management knowledge area. Other processes that belong to this knowledge area include Develop Project Management Plan, Direct and Manage Project Execution, and Monitor and Control Project Work. Estimate Costs is a process that belongs to the Project Cost Management knowledge area and is not an integrative project management process.

   Reference: *PMBOK® Guide,* 4th Ed., pages 71, 73

3. B. A social need, a legal requirement, or a technological advance can be a valid stimulant for chartering a project. Although a political conflict could affect the business of the enterprise, a response to it is usually achieved through business and management strategies, rather than at a project level.

   Reference: *PMBOK® Guide,* 4th Ed., pages 75–76

4. D. Business case, customer request, and contract are all prerequisites before formally chartering and initiating a project. Project management methodology refers to the approach used to manage a project, and does not refer to a prerequisite for formally chartering or initiating a project.

   Reference: *PMBOK® Guide,* 4th Ed., pages 75–76

5. C. Project Purpose or Justification, Assigned Project Manager Responsibility and Authority Level, and Measurable Project Objectives and Related Success Criteria are common items addressed within the project charter, which is produced as part of the Develop Project Charter process. Plan Procurements is a process that belongs to the Project Procurement Management knowledge area and is practiced at later stages of the project after chartering.

   Reference: *PMBOK® Guide,* 4th Ed., pages 76–77

6. A. The SOW incorporates the product scope description, strategic plan, and business need but not the project budget. The project budget establishes the amount of authorized funds for the project. The project scope description describes the characteristics of the product that the project will produce. A business need for an organization may arise due to technological advances, government regulations, industry changes, and legal requirements.

   Reference: *PMBOK® Guide,* 4th Ed., pages 75, 174

7. A. Project Closure Guidelines, Change Control Procedures, and Work Instructions are candidates to be parts of the organization's processes and procedures. They are usable across all projects. The Project Contract, which is a specific document relating to a specific project, is neither a process nor a procedure.

Reference: *PMBOK® Guide,* 4th Ed., page 80

8. C. The outputs of the planning processes address how work will be executed to achieve the project objectives, the project management processes selected by the project management team, and the selected project life cycle. The final product resulting from the project activities is not an output of the project planning processes because no final products are produced at the project planning phase.

Reference: *PMBOK® Guide,* 4th Ed., pages 81–82

9. D. Outputs of the project planning processes include, among other items, how changes will be monitored and controlled, how work will be executed to accomplish project objectives, and the need for techniques of communication among stakeholders. Techniques for negotiating with subcontractors are relevant across many projects. However, these techniques are not included in the project management plan.

Reference: *PMBOK® Guide,* 4th Ed., pages 81–82

10. B. Except for the project charter, all of the other options are part of executing the project management plan. Because no project should start without an authorized project charter, developing and preparing a draft copy and getting a sign-off on the project charter precede project execution.

Reference: *PMBOK® Guide,* 4th Ed., page 93

11. A. Managing project and management contingency reserves is part of cost management. Establishing a change control board that oversees the overall project change environment is an activity of change control, as is influencing factors that cause change and determining change has occurred. The primary objectives of the Perform Integrated Change Control process are to maintain the integrity of baselines, integrate product and project scope, and coordinate changes across knowledge areas.

Reference: *PMBOK® Guide,* 4th Ed., page 93

12. B. Project plan development, overall change control, and project initiation are activities that cover limited aspects of the overall project. They do not involve the whole development teams, and they do not consume the majority of the project budget. The project plan execution requires the project manager and the project team to perform multiple actions to execute the project management plan to accomplish the work defined in the project scope statement.

Reference: *PMBOK® Guide,* 4th Ed., page 83

13. D. Defining key project objectives and documenting project assumptions and constraints are completed prior to developing the project management plan. However, because the planning process is iterative, approved adjustments may be made later on in the project. Promoting communications among stakeholders is done throughout the project execution. The main purpose of developing the project management plan is to create a document to guide project execution and control.

Reference: *PMBOK® Guide,* 4th Ed., page 78

14. C. Completing the project within approved budget and schedule, and ensuring that it satisfies the needs for which it was undertaken, are among the overall objectives of all the Project Integration Management processes and activities. Defining proper scope for the project is only one of the processes of this knowledge area. The collective aim of Project Integration Management processes is to ensure the proper coordination of the various elements of the project.

Reference: *PMBOK® Guide,* 4th Ed., page 71

15. D. Schedule development, collection of reports or tools and techniques to gather, and integrate and disseminate project output are all individual aspects of the PMIS. Automating them individually can hardly be considered as the most relevant to the PMIS. Although theoretically a manual PMIS can be developed for very small projects, for nearly all reasonably sized projects the PMIS must be mostly automated to be effective, and it must be easily usable by the project management team.

Reference: *PMBOK® Guide,* 4th Ed., page 87

16. B. Performance measurement baselines, the change management plan, and the quality assurance plan do not reflect approved changes. The project's change control procedures detail the steps by which official company standards, policies, plans, and procedures, or any project documents, will be modified and how any changes will be approved and validated. For approved changes to be implemented and delivered by the project, they have to be reflected in the project plan. The project plan must be modified to reflect the activities and resources required to implement the approved changes.

Reference: *PMBOK® Guide,* 4th Ed., page 80

17. B. Assumptions are factors that are considered to be true for planning purposes, whereas constraints are factors that will limit the project management team's options. Assumptions are specific to individual projects, and they are not necessarily dependent on either historical information or Lessons Learned. Assumptions are based on factors that may not be true, may not be accurate, or may not be available.

Reference: *PMBOK® Guide,* 4th Ed., page 116

18. C. Expert judgment is the only tool or technique used in the Develop Project Management Plan process. Product analysis, alternatives identification, and facilitated workshops are tools and techniques of the Define Scope process.

Reference: *PMBOK® Guide,* 4th Ed., pages 81, 104

19. D. The integration effort involves making trade-offs among competing objectives and alternatives, which requires communications and negotiation with a variety of stakeholders throughout the project. The sponsor may choose not to have direct communications with the project team. Frequent periodic meetings do not necessarily accelerate the project integration process. Assigning responsibilities to the project team members is an activity for planning the project.

Although integration occurs throughout the project cycle, not just through crisis, crisis periods provide the opportunity to reinforce and accelerate the project integration process by considering the spectrum of opinions and then establishing a consensus on the best course of actions.

Reference: *PMBOK® Guide,* 4th Ed., page 71

20. D. A CCB is necessary for all projects, large and small. A CCB is not typically headed by the project manager, and it is at a higher level than the Configuration Management Board. The CCB is responsible for approving or rejecting changes to the project baselines.

Reference: *PMBOK® Guide,* 4th Ed., page 94

21. C. Neither the general management processes nor the personnel management processes are specific to one project; rather, they are common across all projects in the organization. Also, generally a project would consist of multidisciplinary teams and will not be constrained to follow one discipline-specific process. The project management team follows project management processes, and project members developing the target product or service are expected to follow a set of product-oriented or service-oriented processes.

Reference: *PMBOK® Guide,* 4th Ed., page 37

22. B. The configuration management system includes the processes for identifying the items to be controlled, procedures for submitting and recording proposed changes, a tracking system for reviewing and approving proposed changes, defining approval levels for authorizing changes, providing a method to validate and approve changes, and auditing the controlled items as well as reporting on their status. Changes that might affect management or project reserves are subject to change management. Change management is concerned with changes that affect the project or product baselines.

Reference: *PMBOK® Guide,* 4th Ed., pages 94–95

23. A. The process Direct and Manage Project Execution produces the following outputs: deliverables, work performance information, change requests, project management plan updates, and project document updates.

Reference: *PMBOK® Guide,* 4th Ed., page 73

24. D. Change request status updates, project management plan updates, and project document updates are not outputs of the Perform Integrated Change Control process. Organizational process assets updates are outputs of the Close Project or Phase process and other processes but not the Integrated Change Control process.

Reference: *PMBOK® Guide,* 4th Ed., page 99

25. D. The Collect Requirements process is the process of defining and documenting stakeholders' needs to meet the project's objectives. All of the other options are simply incorrect or inaccurate.

Reference: *PMBOK® Guide,* 4th Ed., page 105

26. C. Sponsor's approval of the project budget, stakeholder identification, and designing the project organization are all activities in the Initiating and Planning phases. As part of directing and managing project execution, the project team performs the work as outlined in the project management plan. Such work entails, among other activities, implementing the planned methods and standards.

Reference: *PMBOK® Guide,* 4th Ed., page 83

27. D. Developing the project management plan, monitoring and controlling the project, and performing integrated change control do not involve producing deliverables. Deliverables, work performance information, change requests, project management plan updates, and project document updates are outputs of the Direct and Manage Project Execution process.

Reference: *PMBOK® Guide,* 4th Ed., page 87

28. C. Requirements are not used as the basis for the project charter, which comes before the requirements. Also, requirements do not have enough details to be suitable for creating work packages for the project, and requirements are not a form of variance analysis for the project. However, product and project requirements, if properly captured with enough detail, become the foundation of the WBS, cost planning, schedule planning, and quality planning.

Reference: *PMBOK® Guide,* 4th Ed., page 105

29. B. Change requests, work performance information, and deliverables are all outputs of the Direct and Manage Project Execution process. Planning processes should be defined prior to the project execution phase.

Reference: *PMBOK® Guide,* 4th Ed., pages 87–88

30. D. Change request status updates are an output of Perform Integrated Change Control; organizational process asset updates are an output of Close Project or Phase as well as other processes but *not* the Direct and Manage Project Execution process; and the project management plan is an output of Develop Project Management Plan. Deliverables are an output of the Direct and Manage Project Execution activities.

Reference: *PMBOK® Guide,* 4th Ed., pages 73, 87

31. C. Performance reports—which include costs incurred, deliverable status, schedule progress, and other project updates—are part of the information collected routinely to measure and monitor work performance. Along with the project management plan, enterprise environmental factors, and organizational process assets, performance reports are inputs. In contrast, schedule forecasts are an output of monitoring and controlling project work.

Reference: *PMBOK® Guide,* 4th Ed., page 87

32. B. Performing activities to accomplish project objectives is part of directing and managing project execution. Collecting and documenting the Lessons Learned is part of the project closure activities, and staffing management and resource planning is part of developing the project management plan. Tracking and monitoring project risks are part of the Monitoring and Control Project process.

Reference: *PMBOK® Guide,* 4th Ed., page 89

33. A. Monitoring implementations of approved changes when and as they occur, comparing actual project performance against the project management plan, and providing forecasts to update current cost and performance against the project management plan are activities in monitoring and controlling the project work. Adapting approved changes into the project scope, plans, and environments is part of directing and managing project execution, not monitoring and controlling activities.

Reference: *PMBOK® Guide,* 4th Ed., pages 83, 89

34. C. Change request status updates are *not* an output of the Develop Project Management Plan process, Develop Project Charter process, or the Direct and Manage Project Execution process. Change request status updates are an output of the Perform Integrated Change Control process.

Reference: *PMBOK® Guide,* 4th Ed., page 99

35. A. Reviewing and approving requested changes, identifying that a change needs to occur or has occurred, and reviewing and approving corrective and preventive actions do not aim mainly at preserving the integrity of project baselines. The integrity of project baselines can be protected by preventing unapproved changes, so releasing only approved changes for incorporation into the project products or services is the main mechanism for preserving the integrity of project and product baselines.

Reference: *PMBOK® Guide,* 4th Ed., page 93

36. D. The main purpose of configuration status accounting is to capture, sort, and report the configuration identification list, proposed changes status, and change implementation status. The basis from which the configuration of products is defined is provided through configuration identification. Establishing that performance and functional requirements have been met is provided through configuration audits, and documenting the technical and functional impacts of requested changes is achieved through configuration control.

Reference: *PMBOK® Guide,* 4th Ed., page 95

37. A. Change request status updates, project management plan updates, and project documents updates are all outputs of the Perform Integrated Change Control process. Work performance information is an input to, and not an output of, the Perform Integrated Change Control process.

Reference: *PMBOK® Guide,* 4th Ed., page 97

38. C. It is true that a CCB is responsible for reviewing change requests and approving or rejecting them, and it is true that the role and responsibilities of the CCB are clearly defined and agreed on by appropriate stakeholders. It is also true that all CCB decisions are documented and communicated to the stakeholders for information and follow-up actions. The statement that the CCB may not necessarily record all of the decisions and recommendations that they have made is incorrect. It is imperative that the CCB accurately and completely document and communicate their work to the appropriate stakeholders.

Reference: *PMBOK® Guide,* 4th Ed., page 98

39. A. Of the available options, the only correct statement is the fact that change requests are approved by the project's CCB. Change requests cannot include corrective actions, preventive actions, and defect repairs because all of these items result in change request implementation. Change requests take place *after* the project initiation, not during project initiation. The word *some* in the last option made it incorrect.

Reference: *PMBOK® Guide,* 4th Ed., pages 97–99

40. C. The statements are correct *except* for deciding whether to accept or reject product or project changes. These changes should occur prior to closing the project and not during the Close Project or Phase process.

Reference: *PMBOK® Guide,* 4th Ed., pages 99–100

41. D. Closing the project or phase includes reviewing all prior information to ensure that all project work is complete, archiving project information for future use by the organization, and investigating and documenting reasons for the project termination. Verifying and deciding on all submitted change requests should have been done through the Perform Integrated Change Control process as part of monitoring and controlling the project.

Reference: *PMBOK® Guide,* 4th Ed., pages 99–100

42. A. The two main categories of organizational process assets (OPAs) are processes and procedures, and corporate knowledge base. Standardized guidelines and work instructions, as well as organization communication requirements, are part of processes and procedures; process measurement databases belong to the corporate knowledge database. Government or industry standards are enterprise environmental factors, and therefore are not considered a part of organizational process assets.

Reference: *PMBOK® Guide,* 4th Ed., pages 14, 32–33

43. C. The project management plan, accepted deliverables, and organizational process assets, such as closure guidelines, are inputs into the Close Project or Phase process. Approved change requests are the only exception to the list of inputs to the Close Project or Phase process because they are inputs into the Direct and Manage Project Execution process.

Reference: *PMBOK® Guide,* 4th Ed., page 100

44. B. This is not an example of the majority method, even though the majority of end users would benefit, according to the sponsor. The majority decision method would entail a group actually participating in the decision-making process, with more than 50% of the group approving the decision. Also it is not an example of unanimity wherein all stakeholders agree. Nor is it a plurality, wherein one choice is preferred by the largest number of votes. Dictatorship is the correct option because the sponsor made the decision by herself.

Reference: *PMBOK® Guide,* 4th Ed., page 108

45. A. As part of the Closing process group, project administrative closure procedures involve archiving project information for future use, gathering Lessons Learned, and validating that completion and exit criteria have been met. Raising problem reports for nonperforming parts of the system is part of the Perform Integrated Change Control process, which belongs to the Monitoring and Controlling process group.

Reference: *PMBOK® Guide,* 4th Ed., page 100

46. B. Focus groups, observations, and change requests are incorrect options. With prototyping, stakeholders can experiment with a model of the final product and refine requirements. Prototyping also supports the concept of progressive elaboration because stakeholders tend to use iterations of working models, experimentation, feedback, and revision.

Reference: *PMBOK® Guide,* 4th Ed., page 109

47. C. Transfer of finished and unfinished deliverables does *not* take place at the end of a key deliverable, at the end of the contract, or as a result of a legal dispute. If the project was terminated prior to completion, the formal documentation indicates why the project was terminated and formalizes procedures for the transfer of the finished and unfinished deliverables of the canceled project to others.

Reference: *PMBOK® Guide,* 4th Ed., page 99

48. B. Closing the project or phase involves collecting and archiving project files, including project management plan, risk and issue register, and quality baselines. All project team communications are collected and archived as part of the project close activities.

Reference: *PMBOK® Guide,* 4th Ed., page 102

49. D. WBS dictionary, project scope statement, and WBS are *not* inputs to the Define Scope process. Requirements documentation is an input to the Define Scope process. Project Scope Management is concerned primarily with defining and controlling what is and is not included in the project. This is mainly defined in the Define Scope process, which focuses on developing a detailed project scope statement as the basis for future project decisions.

Reference: *PMBOK® Guide,* 4th Ed., page 113

50. A. Inputs to the Create WBS process include project scope statement, requirements documentation, and organizational process assets. Therefore, project scope statement is the correct option and the others are incorrect.

Reference: *PMBOK® Guide,* 4th Ed., page 117

51. C. Interviews, facilitated workshops and focus groups are some of the techniques used in the Collect Requirements process. These tools and techniques and others help elicit, analyze, and record the requirements in enough detail to be measured once project execution begins. The development of the initial project management plan baseline is an exception to the list of tools and techniques for the Collect Requirements process. Also, keep in mind that the development of the project management plan is an iterative process, meaning that it occurs before and after the Collect Requirements process.

Reference: *PMBOK® Guide,* 4th Ed., pages 107–109

52. B. Outputs of the Define Scope process include project scope statement and project document updates. therefore, requested changes, issue log, and communications management plan are incorrect options.

Reference: *PMBOK® Guide,* 4th Ed., page 116

53. D. Project charter, project stakeholders, and work breakdown structure are not included in the project scope statement. They are produced either before or after completing the scope definition. The product scope description is included in the detailed project scope statement, together with product acceptance criteria, deliverables, exclusions, and constraints.

Reference: *PMBOK® Guide,* 4th Ed., page 115

54. A. *Component*, *deliverable*, and *work item* are terms that can be used at any level in a WBS. A work package is at the lowest level of the WBS because it represents a manageable work effort.

Reference: *PMBOK® Guide,* 4th Ed., page 118

55. C. Acceptance criteria, product description, and examples of project components are not contained in the WBS dictionary. The WBS dictionary is developed during the project planning and contains data about the WBS components. It contains the code of account identifier, among other things, which is used for unique identification of the WBS components.

Reference: *PMBOK® Guide,* 4th Ed., page 121

56. A. Scope verification is the process of obtaining the stakeholders' formal acceptance of the completed project scope and associated deliverables, whereas quality control is concerned primarily with the correctness of the deliverables and meeting the quality requirements specified for the deliverables.

Reference: *PMBOK® Guide,* 4th Ed., page 123

57. C. Inspections can take the form of any of the verification techniques: audits, reviews, or walkthroughs. Testing is a validation technique as opposed to a verification technique.

Reference: *PMBOK® Guide,* 4th Ed., page 124

58. A. Inputs to the Control Scope process include project management plan, work performance information, requirements documentation, requirements traceability matrix, and organizational process assets. Therefore, project budget, project charter, and project organization are incorrect options.

Reference: *PMBOK® Guide,* 4th Ed., pages 126–127

59. C. Product analysis, expert judgment, and stakeholder analysis are not used in Control Scope. The Control Scope process includes determining the cause of variance relative to the scope baseline and deciding whether corrective action is required. This is achieved through variance analysis techniques.

Reference: *PMBOK® Guide,* 4th Ed., page 127

60. B. Scope control activities do not produce as an output accepted deliverables, project scope management plan, or approved changes. The Control Scope process can generate change requests, which are processed for review and disposition according to the Perform Integrated Change Control process.

Reference: *PMBOK® Guide,* 4th Ed., page 127

61. D. Change control process, project charter, and project scope management plan are not updated as a result of the Control Scope process. The project management plan is correct because scope control could result in changes in the project management plan.

Reference: *PMBOK® Guide,* 4th Ed., page 128

62. C. The Project Time Management knowledge area includes the following six processes: Define Activities, Sequence Activities, Estimate Activity Resources, Estimate Activity Durations, Develop Schedule, and Control Schedule. The other options are incorrect because they belong to the Project Integration Management knowledge area.

Reference: *PMBOK® Guide,* 4th Ed., pages 43, 131

63. C. Develop Schedule, Define Activities, and Sequencing Activities are processes that do not have performance reviews as part of their tools and techniques. Performance reviews are among the tools and techniques of the Control Schedule process because they produce the schedule variance (SV) and schedule performance index (SPI). SV and SPI are indicators of schedule performance.

Reference: *PMBOK® Guide,* 4th Ed., pages 131, 162

64. D. The Define Activities process in Project Time Management uses decomposition, rolling wave planning, templates, and expert judgment to produce activity list, activity attributes, and milestone list as outputs. The other options are incorrect because they are outputs of Sequence Activities, Estimate Activity Resources, and Estimate Activity Durations processes, which use a different set of tools and techniques.

Reference: *PMBOK® Guide,* 4th Ed., page 131

65. A. Rolling wave planning is a form of progressive elaboration in which the work to be accomplished in the near term is planned in detail and future work is planned at a higher level of the WBS, until more information becomes known. It subdivides the project work packages into smaller components called activities.

Reference: *PMBOK® Guide,* 4th Ed., page 135

66. B. By performing a forward pass using critical path method (CPM), the maximum or critical path is revealed as 22 units (B, C, D, F, I, and J). The critical path dictates the shortest duration that a project can be completed. Although the other paths have shorter units, they are not the correct options because the project cannot be considered finished until all of the tasks in the critical path have been completed, as shown in the following diagram.

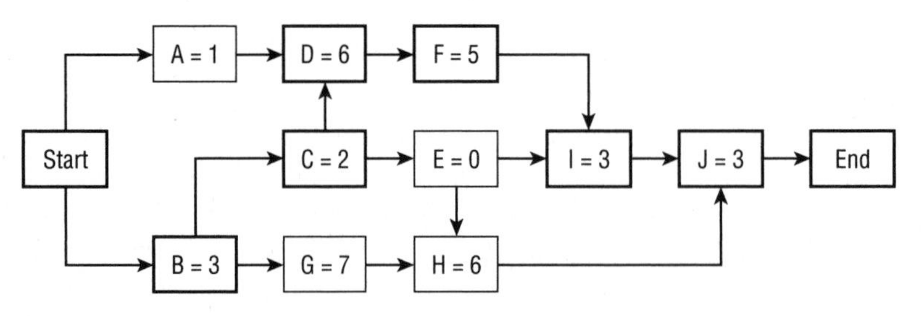

Reference: *PMBOK® Guide,* 4th Ed., pages 138–139

67. C. The critical path is the longest path for example, the path with the longest amount of time. The correct answer is BCDFIJ, as shown in the following diagram.

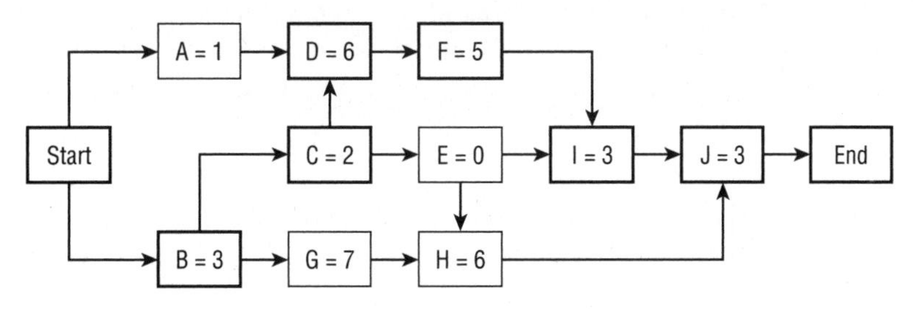

Reference: *PMBOK® Guide,* 4th Ed., pages 138–139

68. D. Estimating activity resources involves determining what resources (persons, equipment, or material) and what quantities of each resource are to be used. Such information is generic and can be delivered by a local industry consultant. The local knowledge of specialized techniques is useful for activity resource estimating, particularly when using expert judgment, alternatives analysis, and published estimating data tools and techniques to produce the outputs. Other options need a knowledge specific to the project, which the local consultant lacks.

Reference: *PMBOK® Guide,* 4th Ed., pages 141–142

69. B. Estimate Activity Duration is the process of approximating the number of work periods needed to complete individual activities with estimated resources. One of the inputs of this process is resource calendars that capture the characteristics (type, quantity, availability, and capability) of both equipment and material resources. Unlike price, make, and supplier, availability is the correct option because it has a direct impact on estimating activity duration.

Reference: *PMBOK® Guide,* 4th Ed., page 148

70. B. The correct names of estimates in the three-point estimating technique are most likely, optimistic, and pessimistic. Least likely is not a type of estimate used in the three-point estimates because it is a made-up term designed to confuse readers by making them think that there is a connection to most likely, which *is* a valid term.

Reference: *PMBOK® Guide,* 4th Ed., page 150

71. C. The purpose of adding duration buffers is not to inflate the project effort, schedule, and plan; not to balance the project resources relative to the project milestones; and not to help manage the total float of network paths. The critical chain method adds duration buffers that are non-work schedule activities, so as to maintain focus on the planned activity durations to help protect the critical chain.

Reference: *PMBOK® Guide,* 4th Ed., page 155

72. A. The use of the term *baseline* in the question hints that the correct name of the output is schedule baseline, which is produced as part of the Develop Schedule process in Project Time Management. Schedule baseline, a component of the project management plan, relies on the schedule network analysis to determine the baseline start and end dates, which are then subsequently reviewed and approved by the project management team. The other options are incorrect because they are inputs, not outputs, of the Develop Schedule process.

Reference: PM *PMBOK® Guide,* 4th Ed., pages 153–154, 159

73. C. Produced by the Direct and Manage Project Execution process as an output, work performance information, along with the project management plan, project schedule, and organizational process assets, are used as inputs to the Control Schedule process. The other options are incorrect because they are outputs, not inputs, of the Control Schedule process.

Reference: *PMBOK® Guide,* 4th Ed., pages 160–161

74. B. The correct option is resource skills. The resource calendar includes availability, capabilities, and skills of human resources. Resource calendars also specify when and for how long project resources will be available during the project, and it may include attributes such as resource experience and/or skill level.

Reference: *PMBOK® Guide,* 4th Ed., page 143

75. C. Root cause analysis is not used in planning the project, estimating the project, or assessing a change request. Root cause analysis focuses on determining the root cause of issues or risks so that efforts can be allocated to address the issues or risks accordingly.

Reference: *PMBOK® Guide,* 4th Ed., page 204

76. B. Resource leveling does not make the schedule shorter, or more responsive to customer needs, or less flexible. Resource leveling is used to address schedule variance activities that need to be performed to meet specific delivery dates; to address the situation in which shared or critical required resources are available only at certain times, or are available only in limited quantities; or to keep selected resource usage at a constant level during specific time periods during the project work. This leveling of resource usage can cause the original critical path to change, usually making the schedule longer.

Reference: *PMBOK® Guide,* 4th Ed., page 156

77. B. EV = Earned Value, SV = Schedule Variance, PV = Planned Value

In the question: PV = 3 weeks, EV = 2 weeks

SV = EV –PV = 2 – 3 = –1 week

Percentage of Variance = Schedule Variance / Planned Value = 1 / 3 = 33%.

The correct option is 33%.

Reference: *PMBOK® Guide,* 4th Ed., pages 162, 182

78. C. EV = Earned Value, SV = Schedule Variance, PV = Planned Value

EV = $850 and PV = $1,000, so SV = EV – PV = –$150.

The correct option is –$150.

Reference: *PMBOK® Guide,* 4th Ed., pages 162, 182

79. D. Direct labor costs, overhead costs, and material costs are inputs to cost budgeting. Sunk costs are costs that have been incurred and that cannot be recovered; therefore, they are not part of the cost budgeting process.

Reference: *PMBOK® Guide,* 4th Ed., page 168

80. C. For month 1, the Cost Performance Index (CPI) = Earned Value / Actual Cost = 1,000,000 / 1,250.000 = 0.88. The statement that CPI at the end of month 1 is 1.25 is not true. A CPI value of less than 1.0 indicates a cost overrun of the estimates, and a CPI value of more than 1 indicates a cost underrun; hence, at the end on Month 1, the project is behind schedule and over budget. At the end of Month 1, Schedule Variance (SV) = Earned Value – Planned Value = $1,000,000 – $1.100.000 = –$100.00, which is true. It is false to state that the CPI at the end of Month 1 is 1.25.

Reference: *PMBOK® Guide,* 4th Ed., pages 181–183

81. A. EV = Earned Value, PV = Planned Value, AC =Actual Cost, CPI = Cost Performance Index (CPI), SV = Schedule Variance

At the end of Month 2, EV is $1,750,000 and AC for the same period is $1,750,000. Thus, at the end on Month 2, CPI = EV / AC = 1,750,000 / 1,750,000 =1; hence the statement that at the end of Month 2 the project is on budget is true.

The statement that at the end on Month 2, the project is behind schedule is not true. The statement that for Month 2, the CPI is 0.66 is not true. The SV for Month 2 = EV – PV = $750,000 – $1,700,000 = $50,000; hence the statement that Month 2 SV is –$300,000 is not true.

Reference: *PMBOK® Guide,* 4th Ed., pages 181–183

82. A. EV = Earned Value; AC = Actual Cost; BAC = Budget at Completion

EAC = AC + BAC – EV = $1,900 + ($1,900 + $1,500) – $1,500 = $3,800

The correct option is $3,800.

Reference: *PMBOK® Guide,* 4th Ed., page 184

83. B. There are several ways to calculate the EAC depending on the approach that the project team would like to use. To calculate the EAC with the assumption that the remaining work will be performed at the same CPI, use the equation EAC = BAC / Cumulative CPI. The equation EAC = AC + Bottom-Up ETC takes into account the current costs plus a new estimate for the remaining work. The equation EAC = AC + BAC – EV uses the actual costs and assumes that the remaining work will be performed according to the budgeted amounts. Lastly, the equation EAC = AC + [(BAC – EV) / Cumulative CPI × Cumulative SPI)] considers both the CPI and schedule performance index (SPI) in the forecast.

Reference: *PMBOK® Guide,* 4th Ed., pages 184–185

84. A. Analogous cost estimating involves using the actual cost of previous, similar projects as the basis for estimating the cost of the current project. It is generally less accurate because there is a limited amount of detailed information about the project (for example, in the early phases). In contrast, parametric provides better estimates compared to analogous especially if the statistical model and the underlying historical data are robust; the three-point estimating technique considers uncertainty and risk by using optimistic, most likely, and pessimistic estimates; and bottom-up estimates can be "rolled-up" for a more accurate estimate.

Reference: *PMBOK® Guide,* 4th Ed., pages 171–172

85. B. The curve illustrates that the value completed is less than planned, so the project is behind schedule, and the actual cost is more than the planned cost, so the project is over budget.

Reference: *PMBOK® Guide,* 4th Ed., page 183

86. C. There is a difference between contingency reserve and management reserve. Contingency reserves are allowances for unplanned, but potentially required, changes. They are included in the cost performance baseline. On the other hand, management reserves are budgets reserved for unplanned changes to project scope and cost. They are not part of the cost performance baseline but are included in the overall budget for the project.

Reference: *PMBOK® Guide,* 4th Ed., page 177

87. B. Quality and grade are not the same. Quality is the degree to which a set of intrinsic characteristics of a product or service are measured to determine whether the requirements were met. Grade is a category assigned to products or services having the same functional use but different technical characteristics. While a quality level that fails to meet quality requirements is always a problem, low grade may not be. Low quality is always a problem; low grade may not be.

Reference: *PMBOK® Guide,* 4th Ed., page 190

88. D. Continuous process improvement can lead to reducing waste, reducing non-value-added activities, and increasing the project efficiency. Continuous process improvement helps to eliminate project risks.

Reference: *PMBOK® Guide,* 4th Ed., page 202

89. C. Quality costs are the total cost of the activities aiming at achieving quality. This includes the cost of preventing nonconformance to requirements, cost of appraising the product or service, and cost of ensuring compliance with mandatory standards. Costs of productivity bonuses are not part of the quality costs. Rather, they are part of the production or development costs. All the other options are part of the quality costs.

Reference: *PMBOK® Guide,* 4th Ed., page 195

90. A. Quality assurance is not about user acceptance testing, applying verification techniques, or planning and implementing inspection. The process of auditing the quality requirements and the results from quality control measurements to ensure appropriate quality standards defines the Perform Quality Assurance process.

Reference: *PMBOK® Guide,* 4th Ed., page 201

91. C. Modern quality management complements project management because both disciplines recognize the importance of customer satisfaction (meeting customer expectations), prevention over inspection (quality should be planned, designed, and built, not inspected), continuous improvement (always look for a better way to do things), and management responsibility (management must provide the resources to succeed). Procurement management is an exception to the aforementioned tenets because it is just one of the nine project management knowledge areas.

Reference: *PMBOK® Guide,* 4th Ed., pages 190–191

92. A. Qualitative and quantitative measures, qualitative assessment of the product or service, and acceptance sampling and statistical process control are techniques used in only one component of quality management. Quality management consists of quality planning, quality control, and quality assurance.

Reference: *PMBOK® Guide,* 4th Ed., page 191

93. C. Project quality assurance does not include policing the conformance of the project team to specifications. It is not a managerial process that defines the organization, design, resources, and objectives of quality management. Also, project quality assurance does not provide the project team and stakeholders with standards by which project performance is measured. Quality assurance provides confidence that the project will satisfy the quality standards that have been defined for the project.

Reference: *PMBOK® Guide,* 4th Ed., pages 201–203

94. B. The project engineer, functional manager, and quality manager do not have ultimate responsibility for quality control. Quality control is the ultimate responsibility of the project manager because it is a project function.

Reference: *PMBOK® Guide,* 4th Ed., pages 206–214

95. B. Project quality audits confirm the implementation of approved change requests and defect repairs, corrective actions, and preventive actions. Management reserve is not part of the cost performance baseline and is not within the authority of the project manager. It is outside the scope of the project quality audit.

Reference: *PMBOK® Guide,* 4th Ed., page 204

96. D. Quality assurance is not concerned with rejected change requests, quality cost report, or testing error reports. Change requests are an important output of performing quality assurance, which is necessary to rectify any noncompliance and to improve the effectiveness and efficiency of the performing organization's policies and processes.

Reference: *PMBOK® Guide,* 4th Ed., page 205

97. A. A Pareto chart is a specific type of histogram, ordered by frequency of occurrence that shows how many defects were generated by type or category of identified cause. Rank ordering is used to focus corrective action so that the effort can be directed to problems that occur the most.

Reference: *PMBOK® Guide,* 4th Ed., page 210

98. C. The Seven Basic Tools of Quality are the cause and effect diagram, control charts, flowcharting, histogram, Pareto chart, run chart, and scatter diagram. A scatter diagram shows the relationship between two variables. A histogram is a vertical bar chart showing how often a particular variable state occurred. A Pareto chart, also referred to as a Pareto diagram, is a specific type of histogram. None of these show the trends of a process over time. A run chart shows the history and pattern of variation over time.

Reference: *PMBOK® Guide,* 4th Ed., page 211

99. C. The Project Human Resource Management knowledge area includes the following processes: Develop Human Resource Plan, Acquire Project Team, Develop Project Team, and Manage Project Team. Administer Payroll is not an official process.

Reference: *PMBOK® Guide,* 4th Ed., page 215

100. B. Project managers use the RACI chart to communicate the roles and expectations within the project. RACI stands for responsible, accountable, consult, and inform. Therefore, assign is not a correct description.

Reference: *PMBOK® Guide,* 4th Ed., page 221

101. A. There are four processes in the Project Human Resource Management knowledge area: Develop Human Resource Plan, Acquire Project Team, Develop Project Team, and Manage Project Team. The other options are incorrect because they are not official processes.

Reference: *PMBOK® Guide,* 4th Ed., page 215

102. B. Activity resource requirements, enterprise environmental factors, and organizational process assets are all inputs of the Develop Human Resource Plan process, which involves human resource planning. The WBS is not an input to Human Resource Planning. It is an output of the Create WBS process in the Project Scope Management knowledge area.

Reference: *PMBOK® Guide,* 4th Ed., page 219

103. A. Project team members' cost, location, skills, and experience are all factors to be considered during project human resource planning. Identifying which departments are involved in the project is not a factor relevant to the project's HR planning as long as there are no issues with availability, costs, and skills.

Reference: *PMBOK® Guide,* 4th Ed., pages 222–225

104. C. The staffing management plan describes when and how the project human resource requirements will be met by incorporating details such as staff acquisition, resource calendars, staff release plan, training needs, recognition and rewards, compliance, and safety. The other options are not part of human resource management.

Reference: *PMBOK® Guide,* 4th Ed., pages 222–225

105. D. Availability, cost, and experience influence the assignment of project team members. Hobbies should not influence the assignment of the project team members.

Reference: *PMBOK® Guide,* 4th Ed., page 227

106. C. Team-building activities, recognition and rewards, and training are all used in team development, which takes place through the Develop Project Team process. Stakeholder analysis is a tool and technique of the Identify Stakeholders process, not the Develop Project Team process.

Reference: *PMBOK® Guide,* 4th Ed., pages 232–234

107. B. The fact that team members socialize after work hours does not mean that they will work more effectively on the project. The other options describe behaviors that indicate effective teamwork.

Reference: *PMBOK® Guide,* 4th Ed., page 230

108. D. Observation, conversation, and project performance appraisal are means for gaining project knowledge in an unplanned training environment. The human resource plan captures scheduled training for the project team, which is a planned training event.

Reference: *PMBOK® Guide,* 4th Ed., page 232

109. A. Team development does not always improve the trust between individual team members. The other options are true about project team development.

Reference: *PMBOK® Guide,* 4th Ed., pages 232–234

110. C. Email communication, a shared goal, and possible inclusion of people with mobility handicaps are all characteristics of virtual teams. Virtual team members are not concerned with managing diverse stakeholder views and expectations. This is the job of the project manager.

Reference: *PMBOK® Guide,* 4th Ed., pages 232–234

111. C. Conflicts often involve ethics, morals, and value systems because many project teams include people from different cultures, with different ethics, morals, and value systems. Interpersonal conflicts produced by these factors are the most difficult to resolve in a project environment.

Reference: *PMBOK® Guide,* 4th Ed., page 239

112. A. Evident improvements in skills, continuous improvements in competencies, and reduced staff turnover are all indicators of team effectiveness. Increased conflict between the project manager and the project personnel is not an indicator of project team effectiveness.

Reference: *PMBOK® Guide,* 4th Ed., pages 229–230

113. A. Understanding compensation plans, benefits, and career paths are the responsibility of functional and personnel managers rather than project managers. The other options describe interpersonal skills necessary for project managers.

Reference: *PMBOK® Guide,* 4th Ed., pages 232, 409–413

114. D. Co-location involves placing many or all of the project team members in the same physical location to enhance their performance as a team. It encourages face-to-face communication, which reduces noise and misunderstandings of the messages communicated. The other options are factors that contribute to effective communications but are not considered to be as effective as co-location.

Reference: *PMBOK® Guide,* 4th Ed., page 234

115. C. Establishing the salary levels of team members is not normally the responsibility of the project manager. Salaries and pay raises are the responsibility of functional and personnel managers, rather than project managers. The other options indicate responsibilities of project managers.

Reference: *PMBOK® Guide,* 4th Ed., page 236

116. D. Tracking team performance, providing feedback, and resolving conflict are all part of the Manage Project Team process. Allocating pay raises is usually the responsibility of a functional manager and not part of project team management.

Reference: *PMBOK® Guide,* 4th Ed., pages 238–241

117. B. Teams will more likely increase effectiveness if they exchange knowledge and experience, share information and resources, and have improved feelings of trust and cohesion among other team members. Encouraging team members to go to lunch together and have fun will not necessarily increase the effectiveness of the project team.

Reference: *PMBOK® Guide,* 4th Ed., pages 229–230

118. A. In projectized organizations, the project manager has full authority over the project resources. Project managers typically hold the view that conflict is inevitable, but managing it properly and resolving it in a constructive manner could lead to positive results for the project.

Reference: *PMBOK® Guide,* 4th Ed., page 239

119. B. The overall level of conflict does not remain constant over the project life cycle, and interpersonal situations are not the most common sources of project conflicts. Also, project managers do not typically favor the compromise approach to handling conflicts. Schedules and priorities are the most common sources of project conflicts.

Reference: *PMBOK® Guide,* 4th Ed., page 239

120. B. Feedback from the project team member's personal friends and family on his or his personality is not part of 360-degree feedback. The other statements are part of the 360-degrees feedback principle.

Reference: *PMBOK® Guide,* 4th Ed., page 238

121. A. Project Communications Management does not cover response planning, stakeholder analysis, or project records and acceptance. Identify Stakeholders, Plan Communications, Distribute Information, Manage Stakeholder Expectations, and Report Performance are major processes of the Project Communications Management knowledge area.

Reference: *PMBOK® Guide,* 4th Ed., page 243

122. C. Plan Communications, Manage Stakeholder Expectations, and Distribute Information are not part of the Monitoring and Controlling process group. Report Performance is the process of collecting and distributing performance information, including status reports, progress measurements, and forecasts, and it is part of the Monitoring and Controlling process group.

Reference: *PMBOK® Guide,* 4th Ed., pages 43, 266

123. D. Noise, message, and encoding are all key components of the basic model of communications. The correct option is messenger because the basic model of communication demonstrates how an idea or information is sent and received between two parties, defined as the sender and receiver. It does not have a messenger as a component of the model.

Reference: *PMBOK® Guide,* 4th Ed., page 255

124. B. Graphical distribution of project staff affects the choice of the communication method and media. Executive requirements affect the contents and frequency of communications. Available technologies affect the method and media of communication. Project length does not affect the communications management plan.

Reference: *PMBOK® Guide,* 4th Ed., pages 253–256

125. D. Presentations, telephone calls, and oral requests are not the best medium for initial response to the client request because they do not formally record the request. Your client's request represents a change to the contract, so the response must come formally, in a letter. This letter will be the formal evidence of the request, which should turn into a change request to the contract.

Reference: *PMBOK® Guide,* 4th Ed., pages 253–256

126. C. The correct answer to the question is not to send him the files because he does not have a legitimate need to know. Information distribution is only to current project stakeholders. Once a member of the team leaves the project, then he or she stops being a legitimate member of the project team.

Reference: *PMBOK® Guide,* 4th Ed., page 258

127. D. The communications management plan includes identifying to whom information will be distributed; what, how, and how often information will be gathered; and what methods will be used to distribute information. Memos, correspondence, and reports from project personnel are actual communication, not part of the project communications management plan itself.

Reference: *PMBOK® Guide,* 4th Ed., pages 256–257

128. C. Stakeholder buy-in, executive interviews, and channel evaluations are not techniques of the Distribute Information process. Communication methods define the techniques used in the information distribution process.

Reference: *PMBOK® Guide,* 4th Ed., page 260

129. B. This activity does not require an update to the project charter, the stakeholder management plan, or the project change control process. The method of information distribution to stakeholders is defined in the communications management plan for the project. This change in information presentation media and frequency must be reflected in updates to the project's communications management plan.

Reference: *PMBOK® Guide,* 4th Ed., pages 256–257

130. B. A request for additional work is effectively a request for change, which should be formally submitted in writing before it is submitted for impact assessment and decision by the appropriate change control authority.

Reference: *PMBOK® Guide,* 4th Ed., page 245

131. A. A simple performance report may only provide a brief status dashboard that focuses on some of the knowledge areas such as scope, schedule, cost, and quality. The results of the ISO audit will not typically be included in a performance report unless that is the only key activity during that reporting period. Even then, the ISO audit results may be lumped in with the quality section of the report and will not be mentioned as a specific item.

Reference: *PMBOK® Guide,* 4th Ed., pages 270–271

132. D. Outputs of the Report Performance process include organizational process assets updates, performance reports, and change requests. Product analysis is not an output of the Report Performance process. It is actually a tool used by the Define Scope process.

Reference: *PMBOK® Guide,* 4th Ed., pages 270–271

133. A. The Identify Stakeholders process is part of the Initiating process group, which is also where the project charter is created. Identifying stakeholders is performed early in the project life cycle, before Planning, Executing, and Monitoring and Controlling.

Reference: *PMBOK® Guide,* 4th Ed., pages 43, 46

134. B. Any request for change should be submitted in written form because a change is likely to have a financial impact and may also have an impact on the contract. There must be formal written evidence of it.

Reference: *PMBOK® Guide,* 4th Ed., page 245

135. C. Performance reporting tools include reporting systems, variance analysis, and status review meetings and are tools and techniques used in the Report Performance process. Project planning sessions are part of the Planning process group and not performance reporting.

Reference: PMBOK® Guide, 4th Ed., pages 268–270

136. B. Project stakeholder management ensures that the communication needs and expectations of various stakeholders are met by proactively soliciting feedback, addressing issues, and clarifying requirements. These are the responsibilities of the project manager. If implemented correctly, project stakeholder management will increase the likelihood of project success and limit the disruptions in the project. It is not true that project team members will spend unnecessary efforts in communicating with key stakeholders.

Reference: *PMBOK® Guide,* 4th Ed., pages 261–262

137. A. Unlike telephone conversations and written or nonverbal communications, verbal face-to-face is the most effective method of communicating complex situations because it allows for immediate feedback and clarification. The sender and receiver of the message can also take advantage of both the verbal and nonverbal (tone of voice, gestures, and so forth) elements of the communication.

Reference: *PMBOK® Guide,* 4th Ed., pages 245 and 264

138. B. The formula for calculating the total number of potential communication channels is [n(n – 1)] / 2, where *n* is the number of stakeholders. So, the correct answer is (8 * 7) / 2 = 28.

Reference: *PMBOK® Guide,* 4th Ed., page 253

139. C. The issue log is not an input to team building, to risk management, or to schedule management. The project's issue log is used to document, monitor, and resolve the project issues in order to maintain good, constructive relationships among the project's stakeholders. The correct option is stakeholder management because the project's issue log is an input to the Manage Stakeholder Expectations process.

Reference: *PMBOK® Guide,* 4th Ed., page 263

140. D. The issue log should contain the follow-on-action agreed on, the root cause analysis of the issue, and the results of any negotiations with functional managers to resolve the issues. The correct option is the results of the project risk analysis because they are not recorded in the issue log but are stored in the risk register.

Reference: *PMBOK® Guide,* 4th Ed., page 263

141. D. Project risks are either known or unknown. Known risks are those that can be identified, analyzed, and planned for. Although it is difficult to prepare a risk response for unknown risks, they should not be ignored. Unknown risks are harder to manage proactively and they cannot be treated as known risks, but they should still be addressed by allocating management reserves.

Reference: *PMBOK® Guide,* 4th Ed., page 177

142. C. Except for the statement that a risk management plan should be updated frequently, the other options are valid statements about risk management plans. Risk management plans need not be updated frequently. They should only be updated if significant changes are made to the project.

Reference: *PMBOK® Guide,* 4th Ed., pages 279–282

143. B. WBS, decomposition, and structured analysis are not used as techniques in the Identify Risks process. The Delphi technique is used when identifying risks as a way to reach a consensus of experts. Risk experts participate in this technique anonymously, helping to reduce bias in the data and keeping any one person from having undue influence on the outcome.

Reference: *PMBOK® Guide,* 4th Ed., page 286

144. C. The risk register is an output of the risk identification activity. It typically contains a list of identified risks, a list of potential responses, and the result of root cause analysis. Work breakdown structure is not contained in the risk register.

Reference: *PMBOK® Guide,* 4th Ed., page 288

145. A. Grouping risks by source of risk, areas of WBS, or project phase could be useful, but they are not correct answers to the question. The correct option is root causes, because grouping risks by common root causes can lead to effective risk responses and better focusing of the effort to address critical areas.

Reference: *PMBOK® Guide,* 4th Ed., page 293

146. B. Sensitivity analysis, a technique used for event- and project-oriented quantitative risk analysis and modeling, typically uses tornado diagrams to compare the relative importance and impact of variables with a high degree of uncertainty against variables that are more stable. Expected monetary value (EMV) analysis uses statistics to calculate the average outcome of scenarios that may or may not happen. Decision tree diagrams are commonly used in EMV analysis. Control charts are used in project quality management, not in project risk management.

Reference: *PMBOK® Guide,* 4th Ed., pages 196, 297–298

147. A. Strategies for dealing with negatives risks include mitigate, avoid, and transfer. Exploit is a strategy for dealing with positive (opportunities) risks and *not* with negative (threats) risks.

Reference: *PMBOK® Guide,* 4th Ed., pages 303–304

148. D. Strategies for dealing with positive risks include exploit, share, enhance, and accept. Increase is not a risk management strategy.

Reference: *PMBOK® Guide,* 4th Ed., pages 304–305

149. B. Updated WBS, change requests, and project organization charts are not inputs to the Monitor and Control Risks process. The project management plan specifies the assignment of risk owners and the resources to project risk management activities.

Reference: *PMBOK® Guide,* 4th Ed., page 309

150. B. Status meetings produce performance reports that are an input to the Monitor and Control Risks process. Variance and trend analysis and technical performance measurement are both techniques used in risk monitoring and control. Project staff rewards scheme is not an input, tool, or technique used in project risk monitoring and control.

Reference: *PMBOK® Guide,* 4th Ed., pages 310–311

151. C. Project Procurement Management processes include Plan Procurements, Conduct Procurements, Administer Procurements, and Close Procurements. The Estimate Costs process is part of the Project Cost Management knowledge area, not Project Procurement Management.

Reference: *PMBOK® Guide,* 4th Ed., page 313

152. D. The procurement management plan, an output of the Plan Procurements process, typically contains the types of contract to be used, standardized procurement documents, and constraints and assumptions. Selected sellers are an output of the Conduct Procurements process, which occurs after the completion of the project management plan.

Reference: *PMBOK® Guide,* 4th Ed., pages 324–325

153. B. The fixed-price contract has the lowest risk for the buyer because the amount to be paid is known and agreed on before the work starts regardless of how the seller will produce the deliverables. Cost-plus-fixed-fee also has a lower risk to the seller because the fee is fixed, but the actual cost of producing the deliverables will not be known until completion. Cost-plus-incentive-fee is similar except that the buyer will need to provide an incentive to the seller depending on their performance. The buyer will have a higher risk with time and materials because there is no incentive for the seller to lower the cost or expedite the production of the deliverables.

Reference: *PMBOK® Guide,* 4th Ed., page 322

154. B. All legal contractual relationships fall into one of two broad families, either fixed-price or cost-reimbursable. Fixed-price contracts are not part of the cost-reimbursable type of contracts.

Reference: *PMBOK® Guide,* 4th Ed., pages 322–324

155. D. Stock price is not a reliable predictor of a seller's ability to deliver the needs of the project. Additionally, if one of the providers is a not-for-profit or private company, this criterion cannot be used. Valid sellers' evaluation criteria may include understanding of the need, technical capability, management approach, technical approach, financial capacity, and other factors.

Reference: *PMBOK® Guide,* 4th Ed., pages 327–328

156. C. Outputs of the Conduct Procurements process include selected sellers, procurement contract award, resource calendars, change requests, project management plan updates, and project document updates. The requested product is not an output of the Conduct Procurements process because it will not be produced until after the seller has been selected.

Reference: *PMBOK® Guide,* 4th Ed., pages 333–335

157. D. Independent estimates, bidder conferences, and procurement negotiation are all tools and techniques used during the Conduct Procurements process. Claims administration takes place during the Administer Procurements process and not during the Conduct Procurements process.

Reference: *PMBOK® Guide,* 4th Ed., pages 331–333

158. D. Plan Procurements, Administer Procurements, and Close Procurements do not have the procurement contract award as an output. Conduct Procurements has procurement contract award as an output along with selected sellers, resource calendars, change requests, project management plan updates, and project document updates.

Reference: *PMBOK® Guide,* 4th Ed., page 333

159. A. Project document updates, resource calendars, and change requests are outputs of the Conduct Procurement process. The procurement statement of work is not an output of the Conduct Procurements process.

Reference: *PMBOK® Guide,* 4th Ed., pages 324, 333–335

160. D. Contract change control systems, inspection and audits, and records management systems are tools or techniques in the Administer Procurement process. Proposal evaluation techniques are used in the Conduct Procurements process and not in the Administer Procurements process.

Reference: *PMBOK® Guide,* 4th Ed., pages 338–339

161. D. Contract discussion during procurement negotiations covers responsibilities and authorities, overall schedule, payment and price, as well as technical solutions. Project personnel is not a subject for contract discussion during procurement negotiations.

Reference: *PMBOK® Guide,* 4th Ed., page 332

162. C. The records management system consists of a specific set of processes, related control functions, and automation tools. The correct option is payment system. Although the records management system may contain records of the actual payments, it does not contain the set of processes for the payment system itself.

Reference: *PMBOK® Guide,* 4th Ed., page 339

163. B. Proposal evaluation techniques and procurement negotiations are other tools and techniques in the Conduct Procurement process, whereas inspections and audits are methods used within the Administer Procurements process. They do not involve real-time meetings with all perspective sellers. Bidder conferences are used to ensure that all prospective buyers have a clear and shared understanding of the procurement process so that there is no preferential treatment.

Reference: *PMBOK® Guide,* 4th Ed., page 331

164. C. As long as the proper change control process is followed and both parties agree on the contract amendments, changes can be introduced into the project. It doesn't matter if changes are introduced at the latter part of the project. There is also no need to consult a neutral organization.

Reference: *PMBOK® Guide,* 4th Ed., page 337

165. A. Plan Procurement, Conduct Procurement, and Administer Procurement are all processes within procurement management. Planning the project takes place earlier than contracting.

Reference: *PMBOK® Guide,* 4th Ed., page 313

166. B. It is true that the contract with selected contractors is subject to remedy in the courts in case of legal disputes. The contract is also a mutually binding legal requirement between the buyer and the seller. The contract obligates the buyer to compensate the seller, and the seller is obligated to provide specified products or services. Take note that contracts can include specific roles and responsibilities.

Reference: *PMBOK® Guide,* 4th Ed., page 333

167. A. Contract administration is usually not part of the project, part of the personnel department, or made external to the organization. Many organizations perform contract administration separate from the project.

Reference: *PMBOK® Guide,* 4th Ed., page 335

168. D. The Conduct Procurements process is used to solicit proposals from prospective sellers. This takes place after the decision to use third-party sellers. Requirements documentation is not a decision process. Procurement audit is a structured review of the procurement process. A make-or-buy decision is used to determine if particular work can be done by the project team or must be purchased from other sources.

Reference: *PMBOK® Guide,* 4th Ed., pages 321, 326

169. C. Conduct Procurement is the process of obtaining seller responses, selecting a seller, and awarding contracts. Inputs to the Conduct Procurement process include organizational process assets, the project management plan, and the qualified sellers list. Selected sellers is an output, not an input of the Conduct Procurements process.

Reference: *PMBOK® Guide,* 4th Ed., pages 330–331

170. B. The Administer Procurements process is about managing procurement relationships, monitoring performance, and making changes and corrections as necessary. These activities interact with those of the following project management processes: Direct and Manage Project Execution, Perform Quality Control, and Perform Integrated Change Control. Initiating the Project is not a process.

Reference: *PMBOK® Guide,* 4th Ed., pages 336–337

171. C. Inputs to the Administer Procurements process include procurement documents, the project management plan, the contract, performance reports, approved change requests, and work performance information. Inspections and audits are tools and techniques, not input, to the Administer Procurements process.

Reference: *PMBOK® Guide,* 4th Ed., pages 337–338

172. D. The procurement performance review is a structured review of the seller's progress in delivering products and services in accordance with contractually stipulated statements of project scope and quality, within cost and on schedule. It is used to identify performance successes or failures, to verify progress in accordance with the procurement statement of work, and to quantify the seller's demonstrated ability to perform work. Selecting the sellers is performed as part of the Conduct Procurements process and not Administer Procurements.

Reference: *PMBOK® Guide,* 4th Ed., page 338

173. B. Close Procurements involves verification of all work and deliverables, updating records to reflect the final results, and archiving the project records. Performing impact assessment of requested changes is part of the change control procedures, which takes place before the Close Procurements process.

Reference: *PMBOK® Guide,* 4th Ed., page 341

174. A. Early termination of a contract may result from mutual agreement between the buyer and seller. The rights and responsibilities in the event of early termination are contained in the contract, and in some cases, the buyer may have to compensate the seller for any accepted work. The correct option is that early termination of a contract may take place only through legal action—this is because early termination can take place by mutual agreement between the buyer and the seller.

Reference: *PMBOK® Guide,* 4th Ed., page 342

175. D. Archive of the procurement file, final acceptance of deliverables, and Lessons Learned could result in updates to the organizational process assets. The correct option is changes to the contract because the contract cannot be updated during the Close Procurements process.

Reference: *PMBOK® Guide,* 4th Ed., page 344

176. C. Contested changes are referred to as claims, disputes, or appeals. The correct option is refund requests because that term is not typically used to refer to contested changes.

Reference: *PMBOK® Guide,* 4th Ed., page 339

177. D. Procurement documentation, an output of the Administer Procurements process, includes the procurement contract, seller-developed technical documentation, seller performance reports, and results of contract-related inspections. The procurement management plan describes how procurement will be managed, including the production of procurement documents, until the closure of the project. Closed procurements are outputs of the Close Procurements process wherein the buyer formally informs the seller that the procurement contract has been completed.

Reference: *PMBOK® Guide,* 4th Ed., page 340

178. D. Tools and techniques for the Close Procurements process include procurement audits, negotiated settlements, and records management system. Expert judgment is a tool or technique of the Close Project or Phase process, among other processes, but not the Close Procurements process.

Reference: *PMBOK® Guide,* 4th Ed., pages 101, 343

179. A. The Close Procurements process is part of the Project Procurement Management knowledge area and its outputs are closed procurements and organizational process assets updates, which include procurement files and Lessons Learned. The project management plan is an input, not an output, of the Close Procurements process.

Reference: *PMBOK® Guide,* 4th Ed., page 344

180. D. Rules and requirements for formal acceptance of the contractual deliverables are not defined in the change control system, the statement of work, or the testing strategy. The correct option is the contract because the requirements and procedures for formal acceptance of the deliverables are defined in the contract.

Reference: *PMBOK® Guide,* 4th Ed., page 344

181. D. The configuration management plan, requirements document, and project plan do not contain the requirements for contract closure. The procurement management plan describes how the procurement process will be managed from developing procurement documents through contract closure.

Reference: *PMBOK® Guide,* 4th Ed., page 344

182. C. Oral communication is not formal enough and cannot be used as an evidence of acceptance or rejection. Written communication may be formal or informal, so it is not reliable as evidence. Change requests are not the right medium to use as an evidence of acceptance or rejection. Formal written notice is considered formal evidence of acceptance or rejection.

Reference: *PMBOK® Guide,* 4th Ed., page 344

183. B. Responsibility, respect, and honesty are not correct options because they, as well as fairness, are the four values identified in the PMI® *Code of Ethics and Professional Conduct.* Although hard work is a desirable trait, it is not one of those values.

Reference: *Project Management Professional (PMP®) Handbook*, page 46

184. C. The *PMI® Code of Ethics and Professional Conduct* applies to all PMI® members as well as individuals who are not members of PMI® but meet one or more of the following criteria: nonmembers who hold a PMI® certification; nonmembers who apply to begin a PMI® certification process; and nonmembers who serve PMI® in a volunteer capacity. The Code does not apply to nonmembers who currently manage projects.

Reference: *Project Management Professional (PMP®) Handbook*, page 46

185. C. The purpose of the *Code of Ethics and Professional Conduct* set of values and standards is to ensure that the project management practitioners do what is right and honorable, set high standards for themselves, and aspire to meet these standards. Applying our standards in a flexible way so that we do not lose our jobs is not part of the Code and often contradicts its principles.

Reference: *Project Management Professional (PMP®) Handbook*, page 46

186. B. To develop the PMI® values, the PMI® used a number of sources, including focus group discussions, two Internet surveys, and management consultants. A direct email survey of 10,000 project managers and projects was not among the PMI® sources of information.

Reference: *Project Management Professional (PMP®) Handbook*, page 50

187. A. According to the PMI®, aspirational standards describe the conduct that we strive to uphold as *practitioners* and adherence to the aspirational standards is not easily measured. Mandatory standards establish firm requirements, and practitioners who do not conduct themselves to these standards will be subject to disciplinary procedures before PMI®'s ethics review committee. Aspirational standards establish firm requirements is not true, and therefore is the correct option.

Reference: *Project Management Professional (PMP®) Handbook*, page 46

188. D. Ignoring this behavior can lead to negative effects on the project. Asking the team leader to stop this behavior is an ineffective resolution, and ignoring this behavior will lead to deterioration of the relationship with the team. Abusive behavior should be recorded and substantiated so that it can be reported in a factual manner to the appropriate functional manager.

Reference: *Project Management Professional (PMP®) Handbook*, page 47

189. C. A member of the project team reporting unethical behavior of another team member or complaining of unethical abuse should not automatically trigger disciplinary action before further investigation. A member of the project team filing an ethics-related complaint without substantiating it with facts does not trigger a disciplinary action without further investigation. When a member of the project retaliates against a colleague who raised an ethical complaint against him or her, a disciplinary action according to PMI® rules is justified.

Reference: *Project Management Professional (PMP®) Handbook*, page 47

190. B. The options of taking on the project and starting self-learning without letting anyone know, or asking your manager not to inform the customer that you do not have the right knowledge are both unethical behavior. Refusing the project without informing your program manager could lead to lost opportunity. Inform the program manager that you do not have experience working in the nuclear industry and that you are prepared to learn on the job. By doing this you fulfill your ethnical obligation and leave him or her to make the decision concerning how to proceed.

Reference: *Project Management Professional (PMP®) Handbook*, page 49

191. C. The correct answer to the question is to ask for permission before copying work products. This corresponds to PMI® guidelines on intellectual property rights. All the other answers contradict the *PMI® Code of Ethics and Professional Conduct* aspirational standard on responsibility that states we protect proprietary or confidential information that has been entrusted to us.

Reference: *Project Management Professional (PMP®) Handbook*, page 47

192. D. Respect/mandatory standard deals with negotiating with others in good faith, not making personal benefits at the expense of others, not acting in an abusive manner toward others, and respecting their property rights. Responsibility/mandatory deals with upholding the policies, rules, regulations, and laws, and reporting unethical or illegal conduct to appropriate management. Responsibility/aspirational deals with basing our decisions and actions on the best interests of society, public safety, and the environment, and our acceptance of assignments consistent with our background, experience, skills, and qualifications. Respect/aspirational deals with our behaviors and attitudes toward others.

The respect/aspirational *PMI® Code of Ethics and Professional Conduct* standard states that "We listen to others' point of view, seeking to understand them." This standard relates to respect for others, and it is aspirational rather than mandatory.

Reference: *Project Management Professional (PMP®) Handbook*, page 48

193. A. Fairness/mandatory deals with conflict of interest situations. Respect/aspirational deals with our behaviors and attitudes toward others. Fairness/aspirational deals with transparency in our decision-making process and objectivity in our actions. The respect/mandatory value standard of the *PMI® Code of Ethics and Professional Conduct* value standard states that "We do not act in an abusive manner toward others." This is mandatory behavior that relates to respect for others; it goes beyond fairness.

Reference: *Project Management Professional (PMP®) Handbook*, page 48

194. A. Responsibility/aspirational deals with basing our decisions and actions on the best interests of society, public safety, and the environment, and our acceptance of assignments consistent with our background, experience, skills, and qualifications. Honesty/mandatory deals with not engaging in behaviors designed to deceive others or in dishonest behaviors. Respect/mandatory deals with negotiating in good faith, seeking to understand others, and conducting ourselves in a professional manner. The fairness/mandatory *PMI® Code of Ethics and Professional Conduct* value standard states that "When we realize that we have a real or potential conflict of interest, we refrain from engaging in the decision-making process." Conflicts of interest must be avoided out of the principle of fairness to all parties involved.

Reference: *Project Management Professional (PMP®) Handbook*, page 48

195. D. Honesty/mandatory deals with not engaging in behaviors designed to deceive others or in dishonest behaviors. Respect/aspirational deals with our behaviors and attitudes toward others. Fairness/aspirational deals with transparency in our decision-making process and objectivity in our actions. Responsibility/mandatory is about upholding the policies, rules, regulations, and laws, and reporting unethical or illegal conduct to appropriate management.

Reference: *Project Management Professional (PMP®) Handbook*, page 47

196. C. Fairness is about acting impartially and objectively. Responsibility is about taking ownership for the decisions we make. Honesty is about understanding the truth and acting in a truthful manner both in our communications and in our conduct. Project managers are expected to uphold the value of respect, that is, to show high regard for one's self, others, and the people who are working on the project directly or indirectly. Safety of others is part of the resources entrusted to us.

Reference: *Project Management Professional (PMP®) Handbook*, pages 47–48

197. D. Responsibility includes taking ownership of our decisions; respect encompasses how we treat ourselves, others, and those people who work in the project; and fairness deals with being impartial in our actions. The correct option is honesty because one of its aspirational standards entails providing information in a timely manner.

Reference: *Project Management Professional (PMP®) Handbook*, page 49

198. D. The *PMI® Code of Ethics and Professional Conduct* states that we proactively and fully disclose any real or potential conflicts of interest to the appropriate stakeholders. As a professional project manager, before making any decisions involving potential conflict of interest you should make a full disclosure to the affected stakeholders, have an approved mitigation plan, and obtain the consent of the stakeholders to proceed. All these actions are at the project level, and it is not necessary to report such a situation to the PMI®.

Reference: *Project Management Professional (PMP®) Handbook*, page 46

199. A. The responsibility/aspirational standard from the *PMI® Code of Ethics and Professional Conduct* states that, "We fulfill commitments that we undertake—we do what we say we will do." Responsibility/mandatory deals with upholding the policies, rules, regulations, and laws, and reporting unethical or illegal conduct to appropriate management. Fairness/aspirational deals with transparency in our decision-making process and objectivity in our actions. Honesty/aspirational deals with seeking to understand the truth, being truthful in our communications and our conduct, and providing accurate information in a timely manner.

Reference: *Project Management Professional (PMP®) Handbook*, page 47

200. B. The responsibility/mandatory standard states that we should only accept assignments where we have the proper background, experience, skills, and qualifications. Fairness/mandatory deals with fully disclosing any real or potential conflict of interest to the appropriate stakeholders, and refraining from attempting to influence outcomes. Respect/aspirational deals with our behaviors and attitudes toward others. Responsibility/aspirational deals with basing our decisions and actions on the best interests of society, public safety, and the environment, and our acceptance of assignments consistent with our background, experience, skills, and qualifications.

Reference: *Project Management Professional (PMP®) Handbook*, page 48

201. D. Given the long duration and the international nature of the project, FP-EPA is the most appropriate contract because it will address the economic fluctuations that may occur during the execution of the project.

Reference: *PMBOK® Guide*, 4th Ed., page 323

202. A. Bid or quotation often applies to procurement where the primary consideration is price. Hence, invitation for bid (IFB) or request for quotation (RFQ) is the correct option.

Reference: *PMBOK® Guide*, 4th Ed., page 326

203. A. The project management plan, approved change requests, enterprise environment factors, and organization process assets are the four inputs of the Direct and Manage Project Execution process.

Reference: *PMBOK® Guide*, 4th Ed., pages 85–86

204. B. Preassignment, negotiation, acquisition, and virtual teams are tools and techniques of the Acquire Project Team process.

Reference: *PMBOK® Guide*, 4th Ed., page 217

205. C. The four stages of a contract life cycle are requirement, requisition, solicitation, and award.

Reference: *PMBOK® Guide*, 4th Ed., page 316

Chapter 9

Practice Test D

1. Three programming activities must be completed on time, in order, to begin testing on time. The first two activities have a 0.90 probability of being completed by the due date. The third has a 0.80 probability of being completed on time. What is the probability of starting the testing on time?

   A. 0.65

   B. 0.70

   C. 0.72

   D. 0.87

2. When does a conflict of interest arise?

   A. If you can acquire more clients because you work on a particular project

   B. If you can realize personal gain because of your professional activities

   C. If you might personally gain because of a decision you make or influence

   D. If your professional activities can favor individuals on the project team

3. What might be allowed if it is the cultural norm in the country where you are working?

   A. Making unauthorized disclosure of confidential or sensitive information

   B. Paying $500 in local currency to get a key shipment released from customs

   C. Representing another company's analytical tool or methodology as your own

   D. Stating that you are experienced in a field when you are not

4. You are working on a project under extreme time pressure that prevents you from spending time on developing a risk management plan. Your boss does not see the value of proper project risk management and he has directed you not to spend any time on this activity. What do you do?

   A. Ask to be moved to a project where risk management is deemed important.

   B. Explain the importance of risk management, try to reach a compromise, and prepare a memo about the meeting; thereafter, do as your boss directs in this matter.

   C. Work overtime and develop a risk management plan anyway.

   D. You have plenty to do as it is; just get on with your assigned tasks.

5. Which of the following terms means the disclosure to others of the unethical behavior of another individual, company, or institution?

   A. Divulgence

   B. Snitching

   C. Squealing

   D. Whistle blowing

6. A supplier in your project provided you with confidential product information that the project team needs in order to integrate the supplier's product into your organization's systems. In the performance of professional services, what are you expected to do with the confidential information?

   A. Always give full disclosure of matters concerning your participation on the project.

   B. Always insist that the customer conform to best project management practices.

   C. Always maintain and respect the confidentiality of sensitive information.

   D. Always maintain and satisfy the scope and objectives of professional services.

7. Which of the following statements is *not* an accurate statement in regard to the PMI® *Code of Ethics and Professional Conduct*?

   A. Be truthful in our communications and in our conduct.

   B. Cooperate with PMI® concerning ethics violations and collection of related information.

   C. Fully disclose any real or potential conflicts of interest to the appropriate stakeholders.

   D. Report any suspected violations of the Code even if we do not have all the facts.

8. The PMI® *Code of Ethics and Professional Conduct* deals with responsibilities, one of the values covered in the code. Which of the following accurately describes the main focus of responsibilities?

   A. Showing high regard to self, others, and resources in the project

   B. Making decisions objectively and acting impartially

   C. Taking ownership of decisions and the consequences

   D. Acting in a truthful manner in our words and actions

9. You are about to make a joint presentation with another company, hoping to obtain a contract worth $250,000. Although you have never worked with this company before, your co-presenter suddenly asks you to say that you have often worked together. What do you do?

   A. Agree because it does not matter anyway

   B. Agree but only if your co-presenter makes the allegation

   C. Refuse because it is not true

   D. Refuse because the companies' demonstrated ability to work together would be a significant advantage from the buyer's perspective

10. Which of the following statements is *not* true of the Develop Project Management Plan process?

    A. It integrates project administration with the organization's administrative procedures.

    B. It integrates project plans and creates a consistent, coherent document.

    C. It integrates the project's work with the organization's ongoing operations.

    D. It provides for the integration of product scope and project scope.

11. Which of the following best describes the total scope of the project?
    A. A collection of work packages and activities planned for the project
    B. A list of requirements the project must address as approved by project owners
    C. Final work breakdown structure
    D. Sum of the project management plan and up to 10 subsidiary plans
12. In planning and managing projects, the project manager must often deal with customer pressures to develop a rich, featured product at low cost, whereas management wants a project that will generate high profits. In dealing with this situation, you should do which of the following?
    A. Brainstorm to find new approaches to providing features at lower cost
    B. Channel customer expectations to a more realistic level
    C. Educate management on a balanced perspective of profit and customer satisfaction
    D. Recognize conflicting objectives among stakeholders and seek a sound compromise
13. What element of the project management plan is essential for progress measurement and control?
    A. Defining required communication tasks
    B. Defining the key management reviews
    C. Documenting project planning assumptions and decisions about chosen alternatives
    D. Providing a baseline
14. Planning meetings and analysis, brainstorming, and stakeholder analysis are used in the Planning process group and are known collectively as which of the following?
    A. High-performance tools
    B. Psychological tools
    C. Structured approach tools
    D. Tools and techniques
15. Which of the following is *not* true about project management information systems (PMIS)?
    A. They are based on electronic databases.
    B. They support push and pull information access methodologies.
    C. They are used to gather, integrate, and disseminate information about the project.
    D. They are used to support all aspects of the project from initiating through closing.

16. Which is correct regarding the project management plan?

A. It is a formal, approved document used to plan, execute, monitor and control, and close the project.

B. It is not complete until all open issues and pending decisions have been resolved.

C. It should be distributed to all stakeholders as soon as it is formally approved.

D. Work breakdown structure items are the key outputs of the project plan development processes.

17. Which of the following are considered part of the performance measurement baseline?

A. Scope, schedule, and cost

B. Scope, schedule, and quality

C. Scope, staffing, and time

D. Time, quality, and risk

18. The ability to sense evolving problems before they can be detected is critical to project success. In the Acquire Project Team process one must therefore pay particular attention to which of the following characteristics?

A. Ability to recognize programming potential

B. Availability, ability, and experience

C. Project management certification

D. Skills that will complement and supplement those of the project manager

19. What is the name for documented recommendations required to bring expected future project performance into conformance with the project management plan?

A. Adjustment action

B. Corrective action

C. Preventive action

D. Response action

20. Which of the following is *not* true concerning status review meetings?

A. The meeting's objectives should include reviewing occurrences since the last meeting and building team spirit.

B. The meeting should provide clear objectives, timeline, and expected deliverables.

C. The meeting should be regularly scheduled.

D. The kind of status review meeting depends on particular stakeholders' needs.

21. Which of the following is an output of the Identify Stakeholders process that is an input to the Plan Communications process?

    A. Communications plan

    B. Project plan

    C. Quality assessments

    D. Stakeholder register

22. Your project budget is $4,750,000. Most of this money will be spent during which of the following?

    A. Project control

    B. Project initiation

    C. Project plan development

    D. Project plan execution

23. Which statement is accurate in regard to a work authorization system?

    A. It is not a part of the overall project management system.

    B. It formally ensures that work is done at the right time in the proper sequence.

    C. Approval levels are not handled as part of the work authorization system.

    D. It should always issue written authorizations with all relevant details.

24. The Direct and Manage Project Execution process entails performing the necessary work to achieve the project's objectives as stipulated in the project management plan. As part of this process, certain approved changes will be incorporated into the project. Such changes include the following *except* ______________.

    A. Corrective actions

    B. Deferred requirements

    C. Preventive actions

    D. Defect repairs

25. A subsidiary plan of the project management plan that may be formal or informal includes details about quality control, quality assurance, and continuous process improvement. What is the name of this plan and which process produces it as an output?

    A. Quality Metrics; Plan Quality process

    B. Quality management plan; Plan Quality process

    C. Change Requests; Perform Quality Assurance process

    D. Quality Control Measurements; Perform Quality Control process

26. The recommended first step in carrying out the Perform Integrated Change Control process is to do which of the following?

A. Add the change request to the agenda for the next review committee meeting

B. Perform a preliminary assessment of the change request

C. Notify the requester that the change request has been received

D. Update the change disposition file

27. You have developed a change control policy. It recognizes that change requests can occur in all the following forms *except* which ones?

A. Externally initiated

B. Informal

C. Optional rather than legally mandated

D. Oral

28. The group that is responsible for approving or rejecting proposed changes has many names. Which of the following is more commonly used?

A. Change assessment board

B. Change control board

C. Engineering assessment board

D. Project control committee

29. Which of the following is *not* an activity of the configuration management system with an integrated change control process?

A. Configuration identification

B. Configuration status accounting

C. Configuration management plan

D. Configuration verification and audit

30. Which change requests can have implications for the project management plans, policies, procedures, schedules, costs, and budgets?

A. Rejected

B. Deferred

C. Declined

D. Approved

31. You are working on a project with multiple key stakeholders who express some conflict with regard to the proper scope of the project. You therefore realize that you must give careful attention to the Perform Integrated Change Control process. What are you concerned with?

- **A.** Carefully managing the primary baselines of scope, time, and cost throughout project execution
- **B.** Creating an effective change control board with well-defined procedures
- **C.** Identifying the precise nature of the stakeholder conflicts and developing appropriately detailed documentation of acceptable compromises
- **D.** The factors that create changes, deciding if a change occurred, and managing changes that do occur

32. The Initiating process group does which of the following?

- **A.** Formalizes the communications and administrative procedures between the customer or end user and the performing organization
- **B.** Obtains authorization to start the project or phase
- **C.** Measures the attractiveness of the proposed project or project continuation to the project owner
- **D.** Progressively elaborates and documents the product description

33. The list of identified project risks and potential responses are documented in the ________________?

- **A.** Risk register
- **B.** Cause and effect diagram
- **C.** System of process flow charts
- **D.** Influence diagrams

34. Which quality planning tool or technique allows for brainstorming of ideas in small groups, and the ideas are then later reviewed by a larger group?

- **A.** Nominal group technique
- **B.** Affinity diagrams
- **C.** Force field analysis
- **D.** Matrix diagrams

35. You have signed a contract with XYZ Software Inc. to develop a software application. You are starting the Initiating processes. In reading the product description you see many areas that are insufficiently described. What should you do next?

A. Ask to be reassigned to a different project where the scope is better defined

B. Document the inadequacies in full detail in a memo that may become the basis for not proceeding with the project

C. Note the inadequacies but do not worry about them because there will be plenty of time to perfect the product description as the project progresses

D. Schedule one or more meetings with the customer to adequately refine the product description so that it can form the basis for the scope definition document

36. Which of the following are *not* inputs to the Initiating process group?

A. Project statement of work and business case

B. Contract and enterprise environmental factors

C. Stakeholder register and stakeholder management strategy

D. Organizational process assets and procurement documents

37. Which of the following represent the five project management process groups?

A. Closing, Initiating, Reporting, Planning, and Assessment

B. Executing, Closing, Initiating, Planning, and Monitoring and Controlling

C. Executing, Planning, Closing, Verification, and Initiating

D. Initiating, Developing, Conceptualizing, Executing, and Closing

38. Which of the following is *not* true about product scope?

A. Completion is measured against the project management plan.

B. It deals with the features and functions that characterize a product or service.

C. Processes, tools, and techniques vary by application area.

D. The product and subproducts are described in the work breakdown structure.

39. The project scope statement describes the deliverables of the project along with the work required to produce those deliverables. It includes details of the following, either directly or via a reference to other documents, *except* _______________.

A. Product scope description

B. Product acceptance criteria

C. Project deliverables

D. Milestone list

40. Who will *least likely* provide a reliable perspective if expert judgment is used in the development of the project scope statement?

A. Consultants

B. Technical associations

C. Industry groups

D. Summer interns

41. Which of the following options best describes the purpose of the Collect Requirements process?

A. Documenting the relationship between the product or service being created and the business need or other stimulus that gave rise to the project

B. Iteratively evolving the work breakdown structure to define the deliverables and subdeliverables at a level of detail where time and cost estimates can be developed reasonably

C. Defining and documenting stakeholders' needs to meet the project objectives

D. Subdividing the major project deliverables into smaller, more manageable components

42. All of the following phrases are source selection criteria *except* _______________.

A. Understanding of need

B. Selecting official biases

C. Technical capability

D. Overall or life-cycle cost

43. Which of the following are assumptions?

A. Factors that, for planning purposes, are considered to be true, real, or certain without proof or demonstration

B. Factors that cannot be verified and so must be considered in the project risk management plan

C. Factors that always should be verified before the final project plan is submitted for approval

D. Factors that will limit the project management team's options

44. Which of the following processes is *not* part of the Project Quality Management knowledge area?

A. Define Quality Metrics

B. Perform Quality Control

C. Perform Quality Assurance

D. Plan Quality

45. Which of the following statements about quality is *false*?

A. Prevention is generally cheaper than inspection.

B. Grade and quality are not the same.

C. Modern quality management and project management are complementary.

D. Precision and accuracy are equivalent.

46. A project team member is performing quality control measurements. If she is using control charts, histogram, run charts, and inspection, which process is she performing?

A. Define Quality Metrics

B. Perform Quality Assurance

C. Plan Quality

D. Perform Quality Control

47. Which of the following statements is *not* true about the project charter?

A. It formally authorizes a project or phase.

B. It should be written by the project manager as one of his first tasks.

C. It should contain the product description.

D. It should include the business need that the project is to address.

48. You are a senior sales manager in a large organization. You have completed your preliminary assessment of a new sales opportunity and have decided to submit an offer. Which of the following should *not* be omitted from your plan to develop the offer?

A. Assemble a team to prepare the offer

B. Determine the project manager who will be responsible for executing the project if the offer is accepted

C. Implement a scope management plan to ensure that the offer is exactly responsive to the prospective customer's expressed needs

D. Write a project charter for the offer development team

49. Which of the following statements applies to qualitative numbers?

A. They are less reliable than quantitative numbers.

B. They are measured by reference to a standard that is external to the system.

C. They cannot meaningfully be added together.

D. They cannot be measured numerically.

50. Which of the following is *not* true about the project scope statement?

 A. It confirms a common understanding among the project stakeholders.

 B. It is an essential component of the project charter.

 C. It outlines the project's product descriptions and specifications.

 D. It provides a documented basis for making future project decisions.

51. Which process performs the subdivision of work into manageable pieces in a deliverable-oriented hierarchical format and defines the entirety of the project as per the approved project scope statement?

 A. Collect Requirements

 B. Define Scope

 C. Create WBS

 D. Verify Scope

52. Which of the following best describes the work breakdown structure?

 A. A deliverable-oriented hierarchical decomposition of the work to be executed by the project team

 B. A product-oriented structure that shows the component elements necessary to produce the project deliverables

 C. An activity-oriented structure that shows which work components have been assigned to which organizational units

 D. An assignment-oriented structure that shows which individuals are responsible for each of the project deliverables

53. Which of the following are the lowest-level items in the work breakdown structure?

 A. Activities

 B. Deliverables

 C. WBS dictionary

 D. Work packages

54. Which of the following is *not* a test element used to verify the correctness of deliverable decomposition in work breakdown structure development?

 A. Are the activities necessary to produce this subdeliverable clearly defined?

 B. Can responsibility for completion of this subdeliverable be assigned to a specific organizational unit?

 C. Considering all the subdeliverables together, are they sufficient for completion of the decomposed item?

 D. Is this subdeliverable necessary for completion of the decomposed item?

55. The work breakdown structure is an output of which planning process?

A. Define Activities

B. Initiate Project

C. Create WBS

D. Plan Scope

56. Systems engineering, value engineering, and function analysis are examples of techniques used in which tool and technique of Define Scope?

A. Alternatives identification

B. Expert judgment

C. Portfolio adjustment

D. Product analysis

57. Which of the following statements is true about the scope management plan?

A. It is also known as the integrated change management plan.

B. It should be an informal written document.

C. It is not used as an input to the Create WBS process.

D. It should include an assessment of the expected stability of the project scope.

58. Which of the following statements is *not* true about the verify scope process?

A. Inspection focuses exclusively on customer deliverables.

B. Inspection takes place when deliverables are completed.

C. The objective is to ensure the correctness of the work results.

D. It is primarily concerned with the acceptance of the work results.

59. Which of the following best describes the purpose of the Verify Scope process?

A. Influencing the factors that create scope changes to ensure that changes are agreed upon, determining that a scope change has occurred, and managing the changes when and if they occur

B. Obtaining formal acceptance of the project scope by the customer or sponsor

C. Progressively elaborating on and documenting the project work that produces the product of the project

D. Subdividing the major product deliverables into smaller, more manageable components

60. Which of the following accurately describes project exclusions?

A. State explicitly what is outside the scope of the project to help manage stakeholder expectations

B. Include the project's product or service, and the associated ancillary reports

C. Set the project scope limits that impact the team's options in regard to schedule, budget, or other factors

D. List various project assumptions that if proven true can impact the outcome of the project

61. The objective of the Define Activities process is to do which of the following?

A. Identify the key activities that must be performed in order to achieve the various project objectives

B. Identify the specific activities that must be performed to produce the various deliverables and subdeliverables

C. Identify the milestones that need to be defined so that the project objectives will be met

D. Identify the specific activity attributes that should be included in the work breakdown structure

62. Which of the following is *not* an input to the Define Activities process?

A. Scope baseline

B. Enterprise environmental factors

C. Templates

D. Organizational process assets

63. Hard logic is also known as which of the following?

A. Discretionary dependencies

B. External dependencies

C. Mandatory dependencies

D. Preferred dependencies

64. Which of the following statements is *not* true about milestones?

A. They correspond to a readiness meeting to move on to the next phase.

B. They correspond to the beginning or end of a significant event.

C. They represent no cost or time expenditures for the project.

D. They should be considered when sequencing activities.

65. The precedence diagramming method (PDM), a method used in critical path methodology (CPM), is used by several software project management software packages. It is also known as ______________.

A. Activity-on-node (AON)

B. Project management information system (PMIS)

C. Schedule performance index (SPI)

D. Budgeted cost of work scheduled (BCWS)

66. Which of the following is *not* a valid time-related constraint?

A. Finish not later than

B. Finish on or before

C. Start on or not earlier than

D. Start two days later than

67. Which statement is accurate in regard to a resource calendar?

A. It contains information about vacation and training time of resources.

B. It deals with the work times applicable for all human and equipment resources.

C. It is a component of the project calendar.

D. It is not appropriate for use with resource pools.

68. The only process that uses decomposition as a tool or technique is the ______________ process.

A. Create WBS

B. Verify Scope

C. Control Scope

D. Develop Project Management Plan

69. You have finished analyzing activity sequences and have determined the start and finish dates for the project activities. What else must you do before the Develop Schedule process is complete?

A. Allocate the available budget to the various tasks

B. Consult with the customer and your top management to secure their approvals

C. Determine the appropriate contingency time to be included in the schedule using PERT techniques

D. Review your schedule in light of available resources and their constraints

70. Tools and techniques in Control Scope include which of the following?

A. Expert judgment

B. Variance analysis

C. Decomposition

D. Inspection

71. Which of the following refers to estimating a task's duration by extrapolating from one or more similar activities completed in the past?

A. Analogous estimating

B. Estimating by extension

C. History-based estimating

D. Parametric estimating

72. Which of the following would normally *not* result in an increased duration for an activity relative to its base estimate?

A. The activity has three people assigned who must all work together to complete the task.

B. The activity involves the performance of detailed and accurate work.

C. The activity requires extensive interaction with various end users.

D. The assigned resource has other concurrent responsibilities and cannot work continuously on the task.

73. Which of the following statements would *not* suggest that the task could be completed sooner by adding more resources?

A. A programming task includes screen development and data modeling.

B. A telephone installer must connect ten new telephone lines into the trunk line array (a 25 × 20-inch metal box).

C. A very small room must be painted.

D. Several experiments must be conducted in which the response of mice to various dosages of a new drug will be assessed.

74. You are in the process of identifying the activities that the project team must perform to produce the deliverables. You are also intending to document the relationships of the activities in addition to trying to estimate the resources for those activities. Which knowledge area are you currently performing?

A. Project Integration Management

B. Project Time Management

C. Project Scope Management

D. Project Quality Management

75. Summary activities are the same as ______________.

A. Fragments

B. Gates

C. Hammocks

D. Nodes

76. Which statement is true regarding the Develop Schedule process?

A. Activity duration estimating is one of the process's central tools and techniques.

B. It must often be iterated along with the processes that provide inputs prior to determination of the project schedule.

C. It should be made by the person or group on the project team who is most familiar with the nature of a specific activity.

D. The network logic diagram is an output of the process.

77. Market demand, organizational need, customer request, technological advance, and legal requirement can result in the creation of the ______________.

A. Business case

B. Contract

C. Project statement of work

D. Enterprise environmental factors

78. The project sponsor mandated that the lease on the office building where most of the project team members perform the work must end on the final day of the Close Project or Phase process. On the project schedule network diagram, the ______________ dependency or logical relationship will be used to represent these activities.

A. Finish-to-start (FS)

B. Finish-to-finish (FF)

C. Start-to-start (SS)

D. Start-to-finish (SF)

79. The predecessor Task A will require seven days to complete and the successor Task B will require nine days to complete. It is desired to start Task B three days before Task A is scheduled to finish. Which of the following dependency relationship representations will produce this result?

A. Task A FF+3 Task B

B. Task A FF-4 Task B

C. Task A FS-2 Task B

D. Task A FS-3 Task B

80. Which of the following statements is *not* true about lag?

- **A.** It can be used with all the dependency relationship types.
- **B.** It is represented in network logic tables and diagrams with a plus (+) sign.
- **C.** The successor task must start later relative to the predecessor task.
- **D.** The successor task should start later relative to the predecessor task, but this can be altered by other aspects of network logic.

81. Lead as a type of dependency adjustment means which of the following?

- **A.** The start of the successor task can be accelerated relative to the predecessor task.
- **B.** The predecessor task must start before the successor task.
- **C.** The successor task can finish later in relationship to the predecessor task.
- **D.** The successor task must start before the predecessor task starts.

82. Which is the most common type of dependency relationship?

- **A.** Finish-to-finish
- **B.** Finish-to-start
- **C.** Start-to-finish
- **D.** Start-to-start

83. A finish-to-finish relationship means which of the following?

- **A.** The predecessor task cannot finish until the successor task has finished.
- **B.** The successor task and the predecessor task must finish at the same time.
- **C.** The successor task cannot finish until the predecessor task has finished.
- **D.** The successor task should not finish earlier than the predecessor task but no logical relationship is implied.

84. Which statement is true concerning the precedence diagramming method?

- **A.** It requires dummy activities to properly represent dependencies.
- **B.** It uses boxes or rectangles to represent activities.
- **C.** It uses directed arrows to indicate activities.
- **D.** Only finish-to-start dependencies can be represented.

85. A foreign construction company working on a road project will need to be familiar with local rules and regulations to minimize estimating mistakes. Likewise, a software development team must be well versed in the latest technologies so that programming activities can be identified correctly and the budget allocated accurately. As such, it is imperative that Estimate Activity Resources process be coordinated with the ______________ process.

A. Estimate Costs

B. Control Schedule

C. Perform Quality Control

D. Administer Procurements

86. Your initial schedule requires more resources in November and January than are available to you. The process of adjusting your schedule to accommodate these constraints is called which of the following?

A. Fast tracking or crashing tasks

B. Resource leveling

C. Resource negotiation

D. Schedule finalization

87. Which is the primary objective of decomposition in the Create WBS process?

A. Consistently reduce the overall project costs

B. Accurately develop flexible project schedules

C. Consistently design and produce quality products

D. Reliably estimate and manage the work packages

88. The cost and accuracy of mathematical models used in the Estimate Activity Durations process does *not* usually depend on which of the following?

A. The accuracy of the historical information used to develop the model

B. The experience of the model developer

C. The scalability of the model's parameters

D. Whether the parameters used are readily quantifiable

89. One output of the Define Activities process is the activity list. What is it?

A. A comprehensive list of all schedule activities required on the project

B. An acceptable substitute for the work breakdown structure when planning time is so limited that some steps must be omitted

C. An input to the Define Scope process

D. A comprehensive list of all activities necessary to produce the various project deliverables

90. Which output of the Define Activity process provides additional information about the various components of an activity?

A. Activity list

B. Milestone list

C. Activity attributes

D. Decomposition

91. All of the following are advantages to using templates to create activity lists *except* ______________.

A. Encouraging creative thinking

B. Indicating required resource skills

C. Reducing risk

D. Saving planning time

92. The outputs from the Develop Schedule process include which of the following?

A. Project schedule, risk management plan updates, and supporting detail

B. Project schedule, schedule management plan, and coding structures

C. Project schedule, schedule baseline, schedule data, and project document updates

D. Project schedule, supporting detail and activity list updates

93. The activity in which the execution of a project is simulated multiple times, using probability distributions for the duration of the individual tasks in order to assess the most likely project duration, is which of the following?

A. Critical path method (CPM)

B. Precedence diagramming method (PDM)

C. Monte Carlo analysis

D. Program evaluation and review technique (PERT)

94. Which statement is true about crashing?

A. It is effective only for tasks on the critical path.

B. It is the preferred approach to dealing with schedule problems because it almost always offers a viable solution.

C. It usually does not increase risk, although it normally increases cost.

D. The method is based on adjusting people resources rather than materials or equipment.

95. Doing activities in parallel that normally would be done sequentially in order to compress the schedule describes which of the following?

A. Accelerating

B. Crashing

C. Fast tracking

D. Objectives management

96. Which graphical representation of the project schedule typically lists the activities on the left, the dates across the top, and the duration of each activity displayed as horizontal bars whose lengths are relative to the duration of the activities?

A. Gantt chart

B. Milestone chart

C. Project network diagram

D. Resource loading chart

97. The schedule for project DEF indicated a project duration of 200 working days. At the end of the 100th working day the SPI stands at 0.909, but the CPI is 1.006. What is most likely true about project DEF?

A. The project is seriously behind schedule and is expected to finish late.

B. The project must realize an SPI of approximately 1.1 over the remaining 100 days if it is to finish on time.

C. The project probably will require 220 working days to complete (200/0.909).

D. The project is performing less work than planned.

98. In performance reports related to schedule performance, which is unique to activities in progress?

A. Actual start date

B. Baseline finish date

C. Baseline start date

D. Forecast remaining time

99. One way to reduce the cost of a software development project is to limit the number of design reviews. This might, however, result in higher operating costs for the department that will maintain the software later on. What is the term often used for this broader view on costing?

A. Activity-based costing (ABC)

B. Life-cycle costing

C. Quality-based costing

D. Total cost control

100. Which plan describes how the project costs will be managed and controlled?

A. Project management plan

B. Cost performance baseline

C. Cost management plan

D. Project funding requirements

101. Cost budgeting is the process of allocating which of the following?

A. The overall cost estimate to individual work activities to establish a cost baseline for measuring project performance

B. The overall cost estimate to individual work activities to establish the basis for the project contingency reserve

C. The project budget to the various identified work activities in order to establish a basis for the project contingency reserve

D. The project budget to the various identified work activities in order to establish a basis for the project estimate

102. Which plan establishes the criteria for planning and controlling project costs, sets the rules for performance management, and defines how risk budgets will be managed?

A. Project management plan

B. Risk management plan

C. Cost management plan

D. Quality management plan

103. During project initiation, you were asked to provide a rough order of magnitude (ROM) estimate. Although the guidelines for ROM may vary depending on the organization, it would reasonable to expect that the estimates will be in the range of ______________.

A. ±50%

B. ±10%

C. ±5%

D. ±3%

104. Which typically is *not* a source of historical information that can be used in cost estimating?

A. Commercial cost-estimating databases

B. Project experience of other firms in the same business area

C. Project files

D. Project team knowledge

105. Failure costs (costs of nonconformance) include all of the following *except* ________________.

A. Costs of quality control (QC)

B. Costs of rework

C. Costs of warranty

D. Loss of reputation

106. What is the relationship between cost estimating and pricing?

A. Price = Cost × (1 + IRR), where IRR is set in the corporate strategic business plan.

B. Price is greater than the cost by an amount that corresponds to the organization's target profit levels in each market segment.

C. The price that an organization can charge for the product or service becomes the basis for project cost estimating.

D. There is no clear relationship.

107. Activity cost estimates are quantitative assessments of which of the following?

A. The expected costs of the resources required to produce the product deliverables, expressed in units of currency

B. The labor and material costs to produce the product deliverables

C. The likely costs of the required resources that are summarized, as required by the organization's chart of accounts

D. The likely costs of the resources required to complete project work

108. Which of the following statements is *not* true about bottom-up estimating?

A. It is not a useful methodology in the early phases of project planning processes.

B. Its accuracy increases if the activities are smaller.

C. Because the cost is largely independent of the desired accuracy, it is the methodology of choice.

D. The work breakdown structure must be accurate and complete before this process can be used.

109. The cost of building a roadway depends on numerous factors, such as the grade of the land, the nature of the soil, and the width and surface material of the roadway. If you evolve a mathematical estimating model, what are these project characteristics called?

A. Analogous quantification

B. Factors

C. Parameters

D. Scalability elements

110. You must provide a cost estimate for a new project that is very similar to two other completed projects. One costs $456,342 and the other costs $437,976. Because this new project is a little larger, you decide to estimate it at $480,000. This is an example of which of the following?

A. Bottom-up estimating

B. Brainstorming estimating

C. Lateral thinking estimating

D. Top-down estimating

111. The code of accounts is an important input to the cost estimating processes because of which of its characteristics?

A. It describes the coding structure used by the performing organization to report financial information in its general ledger.

B. It forms the basis for the control aspects of the cost management plan.

C. It must be included in the supporting detail that is an output of this process.

D. It reflects the financial reporting structure that has been defined in the work breakdown structure.

112. Which must be used to compare actual project results to planned or expected values?

A. Information distribution tools and techniques

B. Performance reviews

C. Trend analysis

D. Variance analysis

113. Which of the following is a Project Scope Management process that involves breaking down a project into smaller, more manageable units?

A. Create WBS

B. Control Scope

C. Define Scope

D. Verify Scope

114. A project manager must have a good understanding of the mechanics of power and politics to be effective. Power includes all the following *except* ____________.

A. Ability to get people to do things that they would not otherwise do

B. Ability to handle power struggles and organizational games

C. Potential to change the course of events and overcome resistance

D. Potential to influence behavior

115. Your project is close to completion when you are informed by the customer that he would like to see an additional feature in the project. You realize that the feature was discussed with the project sponsor in an earlier meeting, and he categorically mentioned that he did not want this feature. What is the best course of action?

A. Add the feature because it will improve customer satisfaction.

B. Add the feature because otherwise the project will not be accepted by the customer.

C. Deny the customer request and ask him to open a change request.

D. Talk with the sponsor about the benefits of having this feature.

116. As a project manager, you must formalize acceptance of the project scope and keep the project focused on the business need that it was undertaken to address. When should this be done?

A. After each phase of the project

B. At the beginning of the project

C. At the end of the project

D. When required by the business sponsor

117. What is the term for quantitative assessments of the likely costs of the resources required to complete project activities?

A. Activity duration estimates

B. Bottom-up estimating

C. Cost estimates

D. Cost management plans

118. As a project nears completion, you notice that the functional managers are more interested in trying to find a new position than in concentrating on work. What is their motivation?

A. Physiological

B. Safety

C. Self-esteem

D. Social

119. ABC Limited is having a bad year. Several of its projects were terminated before any products, services, or results were delivered. Some of its project managers left the company without closing their projects. This affected the project documentation process, and the auditors wrote up the company for incomplete records. The company decides to set up a training program on project report writing to ensure that the same thing does not happen again. Which of the following statements should be part of the scope of the training program?

A. Provide knowledge of archiving processes, documentation methods, and analytical skills

B. Provide knowledge of organization structure and statutory requirements, and impart communication skills

C. Provide knowledge of organization structure, project scope, and deliverables, and impart communication skills

D. Provide knowledge of archiving processes, and impart coordinating and motivating skills

120. Which is an output from the Close Procurements process?

A. Closed procurements

B. Contract

C. Proposal

D. Statement of work update

121. You are the project manager responsible for developing a software application based on customer requirements. The customer requirements change frequently, so you use all the following tools to carry out the Perform Integrated Change Control process *except* ________.

A. Change request status updates

B. Expert judgment

C. Project management information system

D. Project management methodology

122. John is a project manager at a software company. The company has two packages at the beta testing stage. They want advice on which package to launch first. John has received a set of project selection criteria from the client. His next step will be to ________.

A. Do a feasibility study

B. Use a scoring model to choose the best option

C. Document the project selection criteria in the project charter

D. Use cash flow analysis methods to choose the best option

123. The Plan Communications process involves determining the information and communications needs of the stakeholders: who needs what information, when they will need it, how it will be given to them, and by whom. If you are going to start communications planning for your project, what will facilitate this process?

A. Availability of a communications management plan

B. Availability of project records and project reports

C. Compilation of all change requests for the project

D. Knowledge of constraints and assumptions

124. Your goal as a project manager is to manage stakeholders to the best of your ability. If there are conflicts among stakeholders, in favor of whom should they be resolved?

A. All stakeholders

B. Business partner/project sponsor

C. Customer

D. Team members

125. What will determine the tangible and intangible costs and benefits of the project?

A. Benefit-cost analysis

B. Expert judgment

C. Product analysis

D. Work breakdown structure review

126. A parametric estimating technique will be accurate in predicting project costs in all of the following conditions *except* ______________.

A. Historical information available is accurate

B. Model is scalable

C. Parameters are quantifiable

D. If not used with other estimating techniques

127. Your company has an agreement with the labor union that states that workers will not be asked to work more than 45 hours per week. What does this agreement become for the project?

A. Assumption

B. Constraint

C. Predefined criterion

D. Restriction

128. Which statement is *false* regarding procurement negotiations?

A. It includes responsibilities, contract financing, and technical and business management approaches.

B. The main consideration of both buyer and seller should be to maximize monetary return.

C. Price is an important consideration during negotiation.

D. The primary objective should be to build a lasting relationship.

129. What are the processes of the Project Cost Management knowledge area?

A. Develop Schedule

B. Estimate Costs and Perform Qualitative Risk Analysis

C. Estimate Costs, Determine Budget, and Control Costs

D. Perform Qualitative Risk Analysis

130. In schedule development, you create a distribution of probable results for each activity and use it to calculate distribution of probable results for the total project. What is the name of this technique?

A. Fast tracking

B. Mathematical analysis

C. Monte Carlo analysis

D. Resource leveling heuristic

131. Which statement is true about the Close Project or Phase process if a project ended prematurely?

A. It documents reasons for actions taken if a project is terminated before completion.

B. It formalizes acceptance of deliverables.

C. It formalizes acceptance of deliverables and documents deliverables.

D. It verifies and documents deliverables.

132. When your objective is to get a lasting win-win solution to a problem, which conflict resolution technique should you use?

A. Avoiding

B. Confronting

C. Smoothing

D. Withdrawing

133. You are a project manager with a sports agency. You are working with a team of copywriters, artists, and client service staff to "rebrand" a famous hockey player. The team has done similar projects in the past. The project is well under way and you are deciding who will win the Top Performer of the Week Award. Who do you choose?

A. Janet, for staying late and working overtime three nights in a row

B. Larry, for compromising on the family front and putting in extra effort during weekends

C. John, for completing his work and submitting status reports on time, five times in the week

D. Maria, for finishing the work on time this week after being late the last two weeks

134. The hourly wage for semi-skilled workers is $14.00. The annual audit shows fringe benefits cost 30% of basic wages, and overhead costs are 60% of wages plus fringe benefits. What is the loaded hourly wage for a semi-skilled worker?

A. $22.40

B. $26.60

C. $29.12

D. $30.33

135. What should you use to anticipate management's approval of the amount of risk you plan to take on with your chosen project approach?

A. Probability analysis

B. The company's risk tolerance level

C. The Delphi method

D. The Monte Carlo method

136. Which statement is *not* correct about the project charter?

A. It identifies the high-level scope of the project.

B. It identifies resource requirements.

C. It identifies the project manager's authority and responsibility.

D. It often is an internal legal document.

137. As project manager, given the actual cost (AC), earned value (EV), and planned value (PV) in the following options, which of these activities most deserves your immediate attention?

A. AC = $2,400 and EV = $3,000

B. AC = $3,000 and EV = $2,900

C. AC = $3,000 and EV = $3,200

D. PV = $3,000 and EV = $2,900

138. What process group is Verify Scope a part of?

A. Executing

B. Initiating

C. Monitoring and Controlling

D. Planning

139. Your calculations show that you have 300 communication channels. With this many communication channels, how many stakeholders do you have?

A. 5

B. 10

C. 25

D. 30

140. You are managing three small but distinct projects each with 5, 7, and 9 stakeholders, respectively. How many potential communication channels collectively do you need to manage?

A. 21

B. 67

C. 155

D. 315

141. The seller delivers a fixed price plus incentive fee project at a cost of $90,000. The terms of the contract are a ceiling price of $120,000, a target cost of $100,000, a target profit of $10,000, and a target price of $110,000. The share ratio is 70/30. What is the final price (your total reimbursement)?

A. $93,000

B. $96,000

C. $97,000

D. $103,000

142. Your company is running a project in a foreign country and the government is one of the stakeholders. As project manager you have been told that the project has two important constraints: time and environment. You have to keep to the due date and at the same time you must ensure that the environment is protected. In the third month, a stakeholder asks for several changes that you validate through the change management process. However, you discover that one of the changes could increase the toxicity of one of the liquids being released into the ground. If you delay making the changes, you will delay the project and the client asks you to move ahead regardless of the consequences. What must you do?

A. Carry out the validated changes.

B. Get the changes re-validated after highlighting the environmental impact.

C. Inform all stakeholders about the impact and refuse to incorporate the changes.

D. None of the above

143. Given a budget at completion (BAC) of $1,000 with an earned value (EV) of $900 and actual cost (AC) of $950, what is the to-complete performance index (TCPI)?

A. 1.11

B. 2.00

C. 1.005

D. 0.95

144. When is a project considered to be complete?

A. Actual cost (AC) and estimate at completion (EAC) are equal

B. Actual cost (AC), planned value (PV), and estimate at completion (EAC) are equal

C. Earned value (EV) and estimate at completion (EAC) are equal

D. Not enough information to make an assessment

145. Which of the following metrics can be used as part of work performance measurements?

A. Planned vs. actual technical performance

B. Planned schedule performance

C. Actual schedule performance

D. Actual cost performance

146. Debra spent the entire day documenting the key purchasing decisions on her project. She also specified how the procurement process will be managed while asking her assistant to start identifying the potential suppliers to her project. She is currently performing the _______________ process.

A. Conduct Procurements

B. Administer Procurements

C. Plan Procurements

D. Close Procurements

147. An initial assessment of the project indicated that marketplace conditions, suppliers, typical terms and conditions, and unique local requirements can influence the Plan Procurements process. The aforementioned factors are collectively called _______________.

A. Past performance or reputation

B. Internal environmental factors

C. External environmental factors

D. Enterprise environmental factors

148. Which risk response strategy can be used for positive risks or opportunities?

A. Acceptance

B. Enhancement

C. Avoidance

D. Mitigation

149. A fixed-price incentive fee (FPIF) contract has a target cost of $130,000, a target profit of $15,000, a target price of $145,000, a ceiling price of $160,000, and a share ratio of 80/20. The actual cost of the project was $150,000. How much profit does the seller make?

A. $1,000

B. $5,000

C. $10,000

D. $15,000

150. In which of the following processes is the Scope Management plan prepared?

A. Control Scope

B. Develop Project Management Plan

C. Verify Scope

D. Define Scope

151. A company is making a decision regarding which model of equipment to buy for its next project. The equipment has a life of four years. Model A will have a cost of owning of $600 a month. The following information is available for Model B. Its purchase price is $20,000, operational cost is $1,000 per year, meantime between failures is 10 months, and damage per failure is $550. Which of the following statements is true?

A. The monthly cost of owning Model B cannot be determined by the information provided.

B. The monthly cost of owning Model B is approximately $555.

C. The monthly cost of owning Model B is approximately $600.

D. The monthly cost of owning Model B is more than the monthly cost of owning Model A.

152. Which process produces an output that must be updated to reflect the changes in the opportunities and threats in the project, and reviewed when considering risk mitigation costs?

A. Plan Risk Management

B. Perform Qualitative Risk Analysis

C. Plan Risk Responses

D. Identify Risks

153. When there is uncertainty associated with one or more aspects of the project, what is one of the first steps to take?

A. Conduct a needs analysis

B. Plan risk management

C. Increase the estimated cost of the project

D. Revise the project plan

154. Which of the following questions is a functional manager most likely to ask?

A. How much money is available to do the task?

B. When will the task be done?

C. Who will perform the task?

D. Why will the task be done?

155. What is the term for the project manager's ability to influence others using power derived from the same source as reward power?

A. Charter authority

B. De facto authority

C. Penalty power

D. Referent power

156. In which type of organization is team building likely to be most difficult?

A. Cross-functional

B. Functional

C. Matrix

D. Projectized

157. What is the conflict resolution method that should be used when there is a high concern for personal goals and a high concern for relationships?

A. Compromising

B. Forcing

C. Problem solving

D. Smoothing

158. Your client informs you that their company has a problem with the current design specification of your product and how it will interface with existing systems. This could be a major problem for your project. The client is coming to your office in one week to discuss the problem and to see what your team can do to overcome this setback. All previous meetings were informal with this client, but now the client wants a formal meeting. What should you do to prepare for this meeting?

A. Assemble the team and ask them to prepare an agenda for topics to discuss.

B. Make sure that each team member has their assignment in preparation for any handouts and needed plans.

C. Make sure that the team has increased productivity so the client can see how efficiently your team is performing.

D. Update the schedules and assume an active role because the client has not given you the specifics of the potential problem.

159. What is a milestone chart?

A. A chart that shows the project network logic and the project's critical path

B. Another name for a bar chart

C. A project schedule network diagram

D. Similar to a bar chart, but only identifies the scheduled start and completion of major deliverables and key external interfaces

160. What kind of system can be used to establish a negotiating sequence by ranking all proposals by quantifying qualitative data to minimize the effects of personal prejudice in evaluating each proposal?

A. Contract negotiation

B. Expert judgment

C. Screening system

D. Weighting system

161. What is the name for a process of aggregating the estimated costs of individual activities or work packages to establish a cost baseline?

A. Determine Budget

B. Control Costs

C. Estimate Costs

D. Crashing

162. What does the term "progressive elaboration" refer to?

A. Adding detail to the scope as more information becomes known

B. Fully documenting each step of the project

C. Letting team members make scope changes

D. Planning and re-planning as part of the project management methodology

163. Which of the following can be one of the objectives of a feasibility study?

A. To determine if the company has the resources needed to complete the project

B. To determine the deliverables of the project and break them down into activities

C. To examine the project management constraint triangle

D. To map the project to company objectives

164. Which of the following is *not* a common reason for an organization to start a new project?

A. To get a competitive advantage in the market

B. To justify a budget request

C. To meet a customer's needs

D. To meet regulatory requirements

165. During which phase do the project sponsors have the greatest influence on the scope, time, cost, and quality of the project?

A. All phases of the project

B. The Executing phase

C. The Initiating phase

D. The Monitoring and Controlling phase

166. Which of the following is *not* a good reason for the project charter to be signed by an executive?

A. To indicate the relative importance and priority of the project within the organization

B. To provide an authority for the project manager when carrying out project plans and activities

C. To provide greater credibility with people who may be asked to contribute resources or join the project team

D. To find an executive who is not on vacation who can quickly approve the project

167. Which of the following is *not* part of expert judgment?

A. Archived records

B. Contractors

C. International Organization for Standardization

D. Project Management Institute

168. Which of the following is *not* part of the project scope statement?

A. Detailed cost estimates

B. Product acceptance criteria

C. Project assumptions

D. Project constraints

169. Who is responsible for defining the project, developing project plans, and scheduling activities?

A. The idea champion

B. The project manager

C. The sponsor

D. The project management team

170. Who authorizes the resources to be used on a project?

A. The client

B. The idea champion

C. The project manager

D. The sponsor

171. Which is not an organizational process asset (OPA)?

A. Economic conditions

B. Guidelines

C. Historical records

D. Plans, policies, and procedures

172. The CEO has asked you to offer a new software service using satellite imaging. Your company has no previous experience in this area, but you have a degree in electrical engineering and moved right away into the planning for this project. What is the *best* next step for you as a project manager to take?

A. Complete scope, time, and cost estimates.

B. Form a Change Control Board (CCB).

C. Define the scope.

D. Perform a risk analysis.

173. Which of the following is *not* a typical term used to mean inspection?

A. Review

B. Audit

C. Approval

D. Walkthrough

174. Which of the following is *not* an enterprise environmental factor (EEF)?

A. Computers and networks

B. Processes and procedures

C. Economic conditions

D. Mind mapping software

175. Which of the following is *not* an input to the creation of the project charter?

A. Contract (if applicable)

B. Responsibility assignment matrix (RAM)

C. Business case

D. Project statement of work (SOW)

176. Which of the following is *not* a group creativity technique?

A. Brainstorming

B. Nominal group technique

C. Unanimity

D. Affinity diagram

177. Which technique evaluates logical paths of events and future decisions to assess the implications of choosing one option over another?

A. Decision rating chart

B. Decision tree analysis

C. Delphi technique

D. Tree decision analysis

178. Which of the following is *not* part of managing interpersonal relationships?

A. Communication

B. Leadership

C. Problem solving

D. Scope, cost, and time estimating techniques

179. Which of the following methods is used to predict project duration by analyzing the sequence of activities that has the least amount of scheduling flexibility?

A. Critical path method

B. Dependency diagramming

C. Gantt chart

D. Delphi technique

180. What does resource leveling often do?

A. Extend the end date of the project

B. Increase total costs for the project

C. Require fewer resources

D. Require more resources

181. Which of the following is *not* included in Estimate Activity Durations process tools and techniques?

A. Expert judgment

B. Analogous estimating

C. Parametric estimating

D. Schedule network analysis

182. Which of the following is needed to develop a detailed project cost estimate?

A. Cost management plan

B. Procurement management plan

C. Project charter

D. Resource requirements

183. The funding for a project has been reduced. What is the *best* action to take in response to this?

A. Inform the customer of impacts and negotiate a change in scope.

B. Inform the customer that the project will be delayed and adjust resources accordingly.

C. Do only as much work as the new budget permits and document the actions taken.

D. Perform detailed financial analysis and renegotiate for adequate funding.

184. What type of organizational structure usually experiences the most team anxiety at project closure?

A. Functional

B. Projectized

C. Strong matrix

D. Weak matrix

185. Which of the following actions is *not* relevant in reducing the critical path?

A. Adding resources

B. Crashing the time schedule

C. Eliminating free floats

D. Paralleling activities

186. Which estimate will *most likely* accurately reflect the actual cost of the project?

A. Analogous

B. Bottom-up

C. Budget

D. Top-down

187. What process is responsible for developing the project scope statement?

A. Define Scope

B. Initiate Scope

C. Develop Project Management Plan

D. Verify Scope

188. Which of the following conflict resolution techniques is most likely to lead to lasting solutions?

A. Confronting

B. Compromising

C. Forcing

D. Smoothing

189. Team-building activities can be categorized into ______________.

A. Tasks and processes

B. Established goals and negotiated roles

C. Communication and conflict management

D. Motivation and leadership

190. In determining the staffing requirements of a project, what does the project manager need *first*?

A. Activity list

B. Organizational chart

C. Resource breakdown structure (RBS)

D. Responsibility assignment matrix (RAM)

191. What is quality control?

A. Evaluating overall project performance on a regular basis

B. Identifying which quality standards are relevant to the project

C. Monitoring specific project results to determine if they comply with relevant quality standards

D. Taking action to increase the effectiveness and efficiency of the project

192. In a large, complex project, who should have the authority to accept or reject a requested change?

A. Change control board (CCB)

B. Idea champion

C. Project manager

D. Sponsor

193. The CEO of your company wants a monthly schedule update of your project. Which of the following would you send to this executive?

A. Arrow-on-node activity diagram

B. Milestone chart

C. Network precedence diagram

D. PERT chart

194. What are the conflict resolution methods that can be used on a project?

A. Compromising, confronting, directing, and smoothing

B. Confronting, forcing, smoothing, and withdrawing

C. Controlling, directing, forcing, and withdrawing

D. Controlling, forcing, smoothing, and withdrawing

195. Which approach to quality improvement is *least* likely to produce positive results?

A. Continuous process improvement

B. Increased inspection

C. ISO 9000 certification

D. Plan-do-check-act (PDCA) cycle

196. Cost of quality (COQ) includes costs related to which of the following?

A. Changes to requirements

B. Ensuring conformance to requirements

C. Exceeding requirements

D. Performing quality control on the requirements

197. To shorten project duration, resources have been added to tasks on the critical path. What should be reviewed next in the plan?

A. Calculation of a new critical path

B. Lag time on the other paths in the network

C. Longest duration of all tasks remaining in the plan

D. Resources assigned to other tasks on the critical path

198. What do activity duration estimates indicate?

A. How many hours a resource will work on an activity

B. The number of work periods required to complete a scheduled activity

C. When an activity is likely to finish

D. When an activity is likely to start

199. What is the top level of Abraham Maslow's hierarchy of needs?

A. Psychological satisfaction

B. Safety

C. Self-actualization

D. Survival

200. How often should executive management review key project deliverables?

A. Daily

B. Upon request

C. Weekly

D. When completed

201. Given an EV of 375 and a PV of 400, is the project behind schedule?

A. No

B. Yes

C. Depends on the value of the CV

D. Not enough information

202. If the EV is 375 and the AC is 325, then the project's cost performance is ________________.

A. Better than expected

B. On track

C. Unfavorable

D. Still recoverable

203. Helen was asked to draw a Pareto chart to determine the common causes of schedule delays on her project. She identified five items, which she labeled as A, B, C, D, and E, with their corresponding defect frequencies of 100, 350, 200, 180, and 250. In what order should the corrective action be applied?

A. A, C, E, and B

B. A, D, C, and E

C. B, A, E, and D

D. B, E, C, and D

204. Two senior executives are having a disagreement on what should be included in the project scope. Their requirements, if added to the project scope, can lead to complex technical challenges that the project team will need to address. The project manager analyzed the situation, prepared various options, and communicated the options and consequences to the senior executives. After an intense negotiation meeting, both executives left the meeting happy knowing that their concerns have been fully addressed. Which personal skill did the project manager use to resolve the issues?

A. Leadership

B. Political awareness

C. Communication

D. Negotiation

205. ______________ is *not* an input to the Close Project or Phase process.

A. Procurement documentation

B. Project management plan

C. Accepted deliverables

D. Organizational process assets

Chapter 10

Answers to Practice Test D

Answer Key for Practice Test D

1. A
2. A
3. B
4. B
5. A
6. C
7. D
8. C
9. D
10. A
11. C
12. D
13. D
14. D
15. A
16. A
17. A
18. B
19. B
20. A
21. D
22. D
23. B
24. B
25. B
26. B
27. B
28. B
29. C
30. D
31. D
32. B
33. A
34. A
35. C
36. C
37. B
38. A
39. D
40. D
41. C
42. B
43. A
44. A
45. D
46. D
47. B
48. B
49. C
50. B
51. C
52. A
53. D
54. A
55. C
56. D
57. D
58. C
59. B
60. A
61. B
62. C
63. C
64. A
65. A
66. D
67. A
68. A
69. D
70. B
71. A
72. B
73. B
74. B
75. C
76. B
77. A
78. B
79. D
80. D
81. A
82. B
83. C
84. B
85. A
86. B
87. D
88. B
89. A
90. C
91. A
92. C
93. C
94. A
95. C
96. A
97. D
98. D
99. B
100. C
101. A
102. C
103. A
104. B
105. A
106. D
107. D
108. C
109. C
110. D
111. D
112. D
113. A
114. B
115. C
116. A
117. C
118. B
119. B
120. A
121. A
122. C
123. D
124. C
125. A
126. D
127. B
128. B
129. C
130. C
131. A
132. B
133. C
134. C
135. B
136. B
137. A
138. C
139. C
140. B
141. D
142. C
143. B
144. D
145. A
146. C
147. D
148. B
149. C
150. B
151. B
152. D
153. B
154. C
155. C
156. C
157. C
158. D
159. D
160. D
161. A
162. A
163. A
164. B
165. C
166. D
167. A
168. A
169. B
170. D
171. A
172. C
173. C
174. B
175. B
176. C
177. B
178. D
179. A
180. A
181. D
182. D
183. A
184. B
185. C
186. B
187. A
188. A
189. A
190. A
191. C
192. A
193. B
194. B
195. B
196. B
197. A
198. B
199. C
200. D
201. B
202. A
203. D
204. B
205. A

1. A. Probability = product of all the individual probabilities = $0.9 \times 0.9 \times 0.8$.

   Reference: *PMBOK® Guide,* 4th Ed., pages 296–330

2. A. Acquiring more clients because you work on a particular project is a specific example of legitimate high-quality project management performance. Personal gain as the result of a decision you make or influence is also correct, but acquiring more clients is a *more correct* answer because more clients can translate to large financial gains versus personal gain, which can be trivial. Realizing personal gains because of your professional activities is not always true because it is legitimate to acquire a good reputation (personal gain) through consistent high-quality project management performance. Favoring individuals because of your professional responsibilities is not true because it is your job as project manager to select those individuals who can best do the job at hand—your only personal gain is better performance of the project for which you are responsible.

   Reference: *Project Management Professional (PMP®) Handbook*, pages 48, 50

3. B. Paying $500 in local currency to get a key shipment released from customs is allowed if it is the cultural norm in the country where you are working. Only this situation is covered by the proviso unless the other situations conform with applicable laws or customs of the country where project management services are being provided.

   Reference: *Project Management Professional (PMP®) Handbook*, page 48

4. B. In this case, you should explain the importance of risk management and then try to reach a compromise and prepare a memo about the meeting. Thereafter, do as your boss directs in this matter. You have a responsibility to maintain and satisfy the scope and objectives of professional services unless otherwise directed by the customer. Asking to be moved to a project where risk management is deemed important, just getting on with your assigned tasks, and working overtime to develop a risk management plan are incorrect options because they are all moving away from your responsibilities as a professional project manager. Part of being certified as a Project Management Professional (PMP)® is educating the customer on why you think the process is needed. You have a professional obligation to make the project successful despite any roadblocks the customer might place in your way.

   Reference: *Project Management Professional (PMP®) Handbook*, page 47

5. A. Squealing and snitching are more derogatory terms and not generally employed. Divulgence implies disclosure of private or confidential matters not related to unethical behavior. For this question, you are expected to know the definition. However, this is tricky because the definition of "divulgence" is not in the *PMBOK Guide®*.

   Reference: *Project Management Professional (PMP®) Handbook*, pages 48–49

6. C. It is expected that you always maintain and respect the confidentiality of sensitive information. It is good to always provide full disclosure regarding your participation in the project, but it has nothing to do with the question. Likewise, you can recommend that the customer use best management practices but you cannot impose it on them. It is also good to always maintain and satisfy the scope and objectives of professional services, but this option has nothing to do with the question.

Reference: *Project Management Professional (PMP®) Handbook*, page 47

7. D. The PMI® *Code of Ethics and Professional Conduct* suggests that we abstain from accusing others of unethical behavior until we have all the facts. Being truthful in communication and conduct, cooperating with PMI® about ethics violations, and fully disclosing real or potential conflicts of interest to the appropriate stakeholders are covered in the *Code of Ethics and Professional Conduct.*

Reference: *Project Management Professional (PMP®) Handbook*, page 47

8. C. The *Code of Ethics and Professional Conduct* deals with responsibilities by expecting professional project managers to take ownership of their decisions as well as the consequences. The other options are incorrect because they describe the other three values, respectively: respect, fairness, and honesty.

Reference: *Project Management Professional (PMP®) Handbook*, pages 46–47

9. D. You should refuse to for the reason given in option D. The PMI® *Code of Ethics and Professional Conduct* states, "We do not engage in or condone behavior that is designed to deceive others, including but not limited to, making misleading or false statements, stating half-truths, providing information out of context or withholding information that, if known, would render our statements as misleading or incomplete." Option A is clearly in conflict and Option B is not acceptable because of your intent to mislead by silence. Option C could be correct, but Option D is the best answer because it also addresses the materiality of the issue.

Reference: *Project Management Professional (PMP®) Handbook*, page 49

10. A. Integrating project administration with the organization's administrative procedures is not true of the Develop Project Management Plan process. Organizational process assets are one of the inputs to the Develop Project Management Plan process; they function more as constraints on the possible solutions represented by the project management plan.

Reference: *PMBOK® Guide,* 4th Ed., pages 78–79

11. C. The final WBS includes all product and project work that covers the total scope of the project. A collection of work packages and activities planned for the project, without the adjective *all*, may not cover the total scope of the project. A simple list of requirements will not necessarily include the other project management activities that are required to deliver those requirements. a project management plan can have more than 10 subsidiary plans, so that option is incorrect as well.

Reference: *PMBOK® Guide,* 4th Ed., page 121

12. D. Recognizing conflicting objectives among stakeholders and seeking a sound compromise is the most practical and correct solution. The other options assume that only one stakeholder perspective is the *right* one and all others should be required to conform. Brainstorming to find new approaches is, of course, the ideal solution but is seldom practical. Managing stakeholder expectations is challenging because stakeholders often have very different objectives that may come into conflict.

Reference: *PMBOK® Guide,* 4th Ed., pages 23–27

13. D. Documenting project planning assumptions and decisions about chosen alternatives is important to Lessons Learned but not performance reporting. Defining required communication tasks and defining the key management reviews are necessary for this requirement but require the baselines in order to measure actual progress against planned progress. Although several of the responses could be correct under specific circumstances, there is only one best answer.

Reference: *PMBOK® Guide,* 4th Ed., pages 258–261

14. D. Most project planning methodologies make use of a combination of tools and techniques to transform inputs into outputs. Tools are something tangible, such as document templates or software programs, whereas techniques are performed by following a defined systematic procedure. A technique may also use a tool. The other options are incorrect because they are only subsets of the collective term *tools and techniques.*

Reference: *PMBOK® Guide,* 4th Ed., pages 92, 451

15. A. A PMIS is a standardized set of automated tools available within the organization and integrated into a system. A PMIS may be based on electronic databases but that is not always the case, whereas the other options are always true about PMIS.

Reference: *PMBOK® Guide,* 4th Ed., pages 83–87

16. A. The Project Management plan is a formal, approved document used to plan, execute, monitor and control, and close the project. It can be considered complete even though there are open issues and decisions as long as they are documented in the project management log, and accepted as open items as part of the approval process. The project management plan can be distributed in draft form so the project team can provide their feedback. The project management plan is the key output of the project plan development processes and not the WBS items.

Reference: *PMBOK® Guide,* 4th Ed., pages 81–82, 256–257

17. A. The so-called "triple constraints" of scope, time, and cost define the project because any change in one of these variables requires appropriate changes in one or both of the remaining two in order to have a feasible, executable plan. Thus, the corresponding baselines of scope, schedule, and cost become the performance measurement baseline.

Reference: *PMBOK® Guide,* 4th Ed., pages 6–7

18. B. Project managers are selected because of their project management skills. If the team members do not have the necessary *product* skills and knowledge, the team may encounter difficulties in creating the required deliverables. The other options are too specific for particular roles, that is, programming and project management. The Acquire Project Team process deals with acquiring all project team members and not just those two roles.

Reference: *PMBOK® Guide,* 4th Ed., pages 225–226

19. B. Corrective action is anything done to bring expected future project performance in line with the project management plan. As the name implies, preventive action is preventive but the question is looking for something to correct an existing situation. *Adjustment* and *response actions* are made-up (not official) terms.

Reference: *PMBOK® Guide,* 4th Ed., page 92

20. A. In addition to the objectives listed in option A, status review meetings should be used to look at upcoming due dates, to identify possible new risks or looming problems, and to ensure that everything is in place for the work that is to be completed.

Reference: *PMBOK® Guide,* 4th Ed., pages 232, 269, and 338

21. D. The Identify Stakeholders process produces the stakeholder register and stakeholder management strategy as outputs. The stakeholder register, together with stakeholder management strategy, enterprise environmental factors, and organizational process assets, are then used as inputs to the Plan Communications process.

Reference: *PMBOK® Guide,* 4th Ed., pages 246–247, 252

22. D. Direct and Manage Project Execution is the primary process for carrying out the work defined in the project management plan—the vast majority of the project's budget will be expended in performing this process.

Reference: *PMBOK® Guide, 4th,* page 83

23. B. A work authorization system is a formal procedure for sanctioning project work to ensure that work is done at the right time and in the proper sequence. The words *not, always*, and *all* make the other options inaccurate or incorrect.

Reference: *PMBOK® Guide,* 4th Ed., page 452

24. B. The Direct and Manage Project Execution process encompasses several activities such as creating deliverables, generating project data, and issuing change requests. Such change requests, if approved, can be categorized as corrective actions, preventive actions, and defect repairs. Deferred requirements are an exception from the aforementioned categories because they were not approved and will therefore not be incorporated into the project.

Reference: *PMBOK® Guide,* 4th Ed., page 83

25. B. The subsidiary plan of the project management plan that was described in the question is the quality management plan that is produced by the Plan Quality process as an output along with quality metrics, quality checklists, the process improvement plan, and project document updates. Although they belong to the proper processes, the remaining options are incorrect because they were not the output that was described in the question.

Reference: *PMBOK® Guide,* 4th Ed., page 200

26. B. Not all change requests are worth the time and effort to do a complete benefit–cost analysis. There may not be time (because of approaching deadlines) to implement a change, regardless of its merit. The change may already have been considered in a slightly different form. The change may not be consistent with project objectives. A preliminary screening assessment is necessary and is typically performed by the project manager, sometimes with assistance from other project leaders.

Reference: *PMBOK® Guide,* 4th Ed., pages 99–100

27. B. A change request may be initiated or influenced by a party that is external to the project (for example, another department or a parent company). The project may also implement changes that others may consider optional. Oral change requests are also acceptable as long as they get formally documented as defined in the change management process. It is not acceptable to have an informal change request process because it will make it difficult to monitor and control the project scope and other aspects of the project.

Reference: *PMBOK® Guide,* 4th Ed., pages 93–96

28. B. Many change control systems include a group responsible for approving or rejecting proposed changes. Common terms are Change Control Board, Engineering Review Board, Technical Review Board, Technical Assessment Board, and a variety of others. Of these terms, CCB is the only one that was explicitly mentioned in the *PMBOK® Guide.*

Reference: *PMBOK® Guide,* 4th Ed., pages 93–96

29. C. Configuration identification, configuration status, and configuration verification and audit are activities of a configuration management system with an approved change control process. The configuration management plan is not an activity but a subsidiary plan of the project management plan.

Reference: *PMBOK® Guide,* 4th Ed., page 95

30. D. No additional project work will be required for declined, deferred, or rejected change requests. Approved change requests will need to be implemented by the project and that can have an impact on existing project management plans, policies, procedures, schedules, costs, and budgets.

Reference: *PMBOK® Guide,* 4th Ed., page 85

31. D. The Perform Integrated Change Control process is concerned with influencing the factors that create changes to ensure that changes are agreed upon, determining that a change has occurred, and managing the actual changes when and as they occur. The other statements are correct but they are not relevant to the focus of the question: integrated change control.

Reference: *PMBOK® Guide,* 4th Ed., page 93

32. B. The Initiating process group consists of those processes that define a new project or a new phase of an existing project by obtaining authorization to start the program or phase. It is also responsible for the identification of stakeholders. The other options are incorrect because they are not directly related to the two main objectives of the Initiating process group—which are to develop the project charter and to identify the stakeholders.

Reference: *PMBOK® Guide,* 4th Ed., page 44

33. A. The risk register, a key output of the Identify Risks process, contains the list of identified risks and potential responses. All of the other options are incorrect because they are risk diagramming techniques.

Reference: *PMBOK® Guide,* 4th Ed., pages 287–288

34. A. The nominal group technique allows for brainstorming of ideas in small groups and those ideas are later reviewed by a larger group. This technique, along with brainstorming (bringing up various ideas for consideration), affinity diagrams (identifying logical groupings visually), force field analysis (comparing the forces for and against change), matrix diagrams (grouping information and showing relationships), and prioritization matrices (ranking various problems or issues) are used to refine the quality requirements and to prepare effective quality management activities.

Reference: *PMBOK® Guide,* 4th Ed., pages 199–200

35. C. The product description generally will have less detail in early phases and more detail in later ones as the product characteristics are progressively elaborated. Documenting the inadequacies in full detail in a memo that may become the basis for not proceeding with the project is based on a false premise (to kill the project), and scheduling one or more meetings with the customer to adequately refine the product description so that it can form the basis for the project scope statement, while also correct, would be the next step after noting the inadequacies. Not worrying about the inadequacies because there will be plenty of time to perfect the product descriptions as the project progresses is appropriate. Although several responses could be correct under specific circumstances, there is only one best answer.

Reference: *PMBOK® Guide,* 4th Ed., page 114

36. C. There are two processes in the Initiating process group: Develop Project Charter and Identify Stakeholders. These processes use the following as inputs: project statement of work, business case, contract, enterprise environmental factors, organizational process assets, project charter, and procurement documents. Stakeholder register and stakeholder management strategy are not inputs of the Initiating process group; specifically, these are outputs of the Identify Stakeholders process.

Reference: *PMBOK® Guide,* 4th Ed., page 46

37. B. The five project management process groups are Initiating, Planning, Executing, Monitoring and Controlling, and Closing.

Reference: *PMBOK® Guide,* 4th Ed., pages 41–43

38. A. Completion of the project scope is measured against the project management plan, but completion of the product scope is measured against the product requirements.

Reference: *PMBOK® Guide,* 4th Ed., page 103

39. D. Either directly or via reference to other documents, the project scope statement includes product scope description, product acceptance criteria, project deliverables, project exclusions, project constraints, and project assumptions. The milestone list, an output of the Define Activity process, is not part of the project scope statement.

Reference: *PMBOK® Guide,* 4th Ed., pages 115–116, 133

40. D. Expert judgment relies on certain expertise gained from years of professional experience in a particular field. Other units within the organization, consultants, customers, professional and technical associations, industry groups, and subject matter experts can provide a more reliable perspective than summer interns.

Reference: *PMBOK® Guide,* 4th Ed., page 114

41. C. Collect Requirements is the process of defining and documenting stakeholders' needs to meet the project objectives. The options iteratively evolving the work breakdown structure to define the deliverables and subdividing the major project deliverables both use the definition of decomposition. Documenting the relationship between the product and service being created involves alignment.

Reference: *PMBOK® Guide,* 4th Ed., pages 105–106

42. B. "Selecting official biases" is a made-up phrase and it is used simply as a distracter. Source selection criteria includes understanding of need, overall or life-cycle cost, technical capability, risk, management approach, technical approach, warranty, financial capacity, production capacity and interest, business size and type, past performance of sellers, references, intellectual property rights and property rights.

Reference: *PMBOK® Guide,* 4th Ed., pages 327–328

43. A. Assumptions are factors that, for planning purposes, are considered to be true, real, or certain without proof or demonstration. Factors that cannot be verified should be captured as risks. Factors that limit the project management team's options are called restrictions, not assumptions.

Reference: *PMBOK® Guide,* 4th Ed., page 427

44. A. The Project Quality Management knowledge area includes three processes: Plan Quality, Perform Quality Assurance, and Perform Quality Control. Although quality metrics is a valid output of the Plan Quality process, Define Quality Metrics is a fabricated process name.

Reference: *PMBOK® Guide,* 4th Ed., page 189

45. D. According to the *PMBOK® Guide*, "Precision and accuracy are **not** equivalent." Repeated measurement of the former has very little scatter whereas the latter's measured value is very close to the target value. It is generally accepted that it is often cheaper to prevent the causes of defects instead of fixing them after inspection. Quality captures the characteristics of the products and how they meet the requirements, whereas grade is a category assigned to products that perform the same functional use based on different technical characteristics. Modern quality management and project management are complementary because they both subscribe to the same principles of customer satisfaction, prevention over inspection, continuous improvement, and management responsibility.

Reference: *PMBOK® Guide,* 4th Ed., page 190

46. D. Although quality metrics is a valid output of the Plan Quality process, Define Quality Metrics is a fabricated process name. Quality control measurements are an output of the Perform Quality Control process. This is further reinforced by the fact that she used control charts, run charts, and inspection to perform the process. All of these are tools and techniques of the Perform Quality Control process, along with cause and effect diagrams, flowcharting, Pareto charts, scatter diagrams, statistical sampling, and approved change requests reviews.

Reference: *PMBOK® Guide,* 4th Ed., page 191

47. B. The project charter should be issued by a manager external to the project and at a level appropriate to the needs of the project. While the project manager may be tasked with writing or assisting in writing the project charter, this is not always the case.

Reference: *PMBOK® Guide,* 4th Ed., pages 73–74

48. B. The project manager should be identified and assigned as early in the project as is feasible. The project manager preferably should be assigned before much project planning has been done. A preassignment is acceptable and may be included as part of a contract. This information would then be transferred to the project charter, once it is written.

Reference: *PMBOK® Guide,* 4th Ed., pages 73–74

49. C. Qualitative numbers may also be measured numerically, so the primary distinction between qualitative and quantitative numbers is that qualitative numbers cannot meaningfully be added together. For example, if two one-gallon pails of water are poured into a two-gallon container, the density of water in the two-gallon container remains at 1.0. Option A could also be a correct option, if not the best answer. Qualitative information is always more subjective than quantitative information, making it less reliable.

Reference: *PMBOK® Guide,* 4th Ed., pages 189–191

50. B. The scope statement provides a documented basis for making future project decisions and for confirming or developing common understanding of project scope among the stakeholders. It should contain a brief summary of the product description and its specifications. It is developed *after* the project charter is created.

Reference: *PMBOK® Guide,* 4th Ed., page 115

51. C. As part of the Project Scope Management knowledge area, the Create WBS process subdivides the project work as per the approved project scope statement into manageable pieces. The WBS is presented in a deliverable-oriented hierarchical format. The other options are incorrect because the Collect Requirements process captures the stakeholders' needs in order to achieve the project objectives; Define Scope describes the project and product in detail; and Verify Scope formalizes the acceptance of deliverables.

Reference: *PMBOK® Guide,* 4th Ed., pages 103, 116

52. A. A work breakdown structure (WBS) is a deliverable-oriented grouping of project components that organizes and defines the total scope of the project; work not in the work breakdown structure is considered to be outside the scope of the project. The other options are incorrect because they are made-up descriptions that do not accurately describe the WBS.

Reference: *PMBOK® Guide,* 4th Ed., page 121

53. D. The items at the lowest level of the WBS are referred to as work packages. The other options are incorrect because activities are work components that must be performed to deliver the work package; deliverables are verifiable products, services, or results that may include multiple work packages; and the WBS dictionary describes each component in the WBS.

Reference: *PMBOK® Guide,* 4th Ed., pages 119, 133, 426, 432, and 453

54. A. The test element that checks for activities is inaccurate because the focus of the question is on the correctness of the deliverable decomposition, not the details of the work required to deliver them.

The criteria for verifying the correctness of the decomposition are as follows:

1. Are the lower-level items both necessary and sufficient for completion of the decomposed item?
2. Is each item clearly and completely defined?
3. Can each item be appropriately scheduled, budgeted, and assigned to a specific organizational unit that will accept responsibility for satisfactory completion?

Reference: *PMBOK® Guide,* 4th Ed., page 133

55. C. The outputs from the Create WBS process are the WBS, WBS dictionary, scope baseline, and project document updates. The other options are incorrect because they do not produce the WBS as an output.

Reference: *PMBOK® Guide,* 4th Ed., pages 121–122

56. D. Product analysis includes techniques such as product breakdown analysis, systems engineering, value engineering, value analysis, function analysis, and quality function deployment. Alternatives identification includes techniques such as brainstorming, lateral thinking, pairwise comparisons and so on; expert judgment leverages the expertise of subject matter experts; and portfolio adjustment is a made-up tool or technique.

Reference: *PMBOK® Guide,* 4th Ed., page 114

57. D. The scope management plan should include an assessment of the expected stability of the project scope and may be formal or informal, highly detailed, or broadly framed, based on the needs of the project. The scope management plan is not also known as the integrated change management plan. It should be a formal written document because it is used as one of the inputs of the Create WBS process.

Reference: *PMBOK® Guide,* 4th Ed., pages 112–113

58. C. Verify Scope differs from Perform Quality Control in that it is primarily concerned with *acceptance* of the work results, whereas Perform Quality Control is primarily concerned with the *correctness* of the work results.

Reference: *PMBOK® Guide,* 4th Ed., page 123

59. B. Verify Scope is the process of obtaining formal acceptance of the project scope by the customer or sponsor. The other options are incorrect because they describe the Control Scope, Define Activities, and Create WBS processes, respectively.

Reference: *PMBOK® Guide,* 4th Ed., pages 116, 123, 125, and 133

60. A. Project exclusions focus on documenting items that are outside the project scope in order to properly manage stakeholder expectations. The other options are incorrect because they describe project deliverables, project constraints, and project assumptions, respectively.

Reference: *PMBOK® Guide,* 4th Ed., pages 115–116

61. B. The Define Activities process specifies the activities that must be performed to produce the project deliverables. The other descriptions, which limit the activities to key activities, activity attributes, and milestones, are incomplete and inaccurate, and are therefore incorrect options.

Reference: *PMBOK® Guide,* 4th Ed., page 133

62. C. The inputs to the Define Activities process are scope baseline, enterprise environmental factors, and organizational process assets.

Reference: *PMBOK® Guide,* 4th Ed., page 134

63. C. Mandatory dependencies are those that are contractually required or inherent in the nature of the work being done. They are also called hard logic. Discretionary dependencies are the same as preferred dependencies. External dependencies capture the dependencies between project and nonproject activities.

Reference: *PMBOK® Guide,* 4th Ed., pages 139–140

64. A. Milestone events, which are significant points in the project, need to be part of the Develop Schedule process to assure that the requirements will be met before moving on to the next activities.

Reference: *PMBOK® Guide,* 4th Ed., page 438

65. A. AON is another name for PDM. A PMIS is a standardized set of automated tools available within the organization and integrated into a system. SPI and BCWS (now referred to as PV, or planned value) are values used in earned value management (EVM) calculations.

Reference: *PMBOK® Guide,* 4th Ed., page 138

66. D. Time-related constraints, those which are tied to a specific date, are Start (no earlier than, on, no later than) and Finish (no earlier than, on, no later than). A lag (delaying the successor activity, for example, start two days later) is tied to the predecessor activity and not on a specific date.

Reference: *PMBOK® Guide,* 4th Ed., pages 138, 437

67. A. Option B, which deals with the work times applicable for all human and equipment resources, refers to *project* calendars. Resource calendars override project calendars in regard to resource availability (resource pools) for such things as vacation and training time.

Reference: *PMBOK® Guide,* 4th Ed., page 143

68. A. The Create WBS process uses decomposition as a tool or technique whereas Verify Scope uses inspection, Control Scope uses variance analysis, and Develop Project Management Plan uses expert judgment.

Reference: *PMBOK® Guide,* 4th Ed., pages 73, 104, 118, and 127

69. D. The first iteration of the project schedule based only on dependency relationships and activity durations indicates only the most optimistic schedule, assuming unlimited resources. Because this is rarely the case, the next step is to determine resource availability and the impact of external constraints so that the schedule can be adjusted appropriately.

Reference: *PMBOK® Guide,* 4th Ed., pages 153–154

70. B. The *only* tool and technique in the Control Scope process is variance analysis whereas the Create WBS process uses decomposition, Verify Scope uses inspection, and Develop Project Management Plan uses expert judgment.

Reference: *PMBOK® Guide,* 4th Ed., page 127

71. A. Analogous estimating, also called top-down estimating, means using the actual duration of a previous similar activity as the basis for estimating the duration of an activity for the existing project. *Estimating by extension* and *history-based estimating* are made-up terms. Parametric estimating incorporates the relationships between historical data and parameters (for example, size, length, weight, and so forth) to arrive at the estimates.

Reference: *PMBOK® Guide,* 4th Ed., page 149

72. B. Although not stated explicitly, this question asks you to select the *best* answer, not one that is necessarily always true. Presumably the nature of the detailed and accurate work would be known at the time the base estimate was made. All the other options involve the uncertainties of coordinating with other team members, which is inherently uncertain.

Reference: *PMBOK® Guide,* 4th Ed., page 145

73. B. A great example to remember for the inability to divide the work between additional resources is that nine pregnant women cannot bear a child in one month. Screen development and data modeling are clearly distinct. One painter can paint two walls and another painter can paint the other two walls. Different researchers can each be responsible for one of the experiments. However, it is not practical to have more than one person working in a 25 × 20-inch area. Therefore, it will be difficult to shorten the duration of this task even if additional resources are available.

Reference: *PMBOK® Guide,* 4th Ed., pages 156–157

74. B. The processes described in the question are Define Activities, Sequence Activities, and Estimate Activity Resources. These processes are part of the Project Time Management knowledge area.

Reference: *PMBOK® Guide,* 4th Ed., page 129

75. C. A hammock is an aggregate or summary activity (a group of related activities is shown as one and reported at a summary level). The summary activities or tasks in some project management software are represented by two diamond endpoints joined by a line. It looks like a hammock, especially to an overworked project manager! The other options are incorrect because a fragment is a subdivision of a project schedule network diagram, gates signify the end phases when phases are performed sequentially, and nodes are boxes or rectangles that represent activities in a PDM.

Reference: *PMBOK® Guide,* 4th Ed., pages 138, 141, 157, and 436

76. B. The Develop Schedule process analyzes activity sequences, durations, constraints, and other factors before the schedule can be finalized. Given this, schedule development is an iterative process. The other options are incorrect because they are simply false statements.

Reference: *PMBOK® Guide,* 4th Ed., page 152

77. A. The business case helps determine whether it is worthwhile to invest the necessary resources in a project. It can be triggered by one or more of the following: market demand, organizational need, customer request, technological advance, legal requirement, ecological impacts, and social need. The other options are incorrect although similar to the business case; they are inputs to the Develop Project Charter process.

Reference: *PMBOK® Guide,* 4th Ed., pages 75–76

78. B. There are four types of dependency or logical relationship, all with names that are self-explanatory that describe the start and finish connections between two activities: FS, FF, SS, and SF. Because the two activities in the question must finish at the same time, the correct answer is FF.

Reference: *PMBOK® Guide,* 4th Ed., page 138

79. D. Based on the question, it is clear that Task A must finish before Task B can start so this can be achieved by a finish-to-start (FS) dependency: Task A FS Task B. However, there is an extra constraint that Task B must start three days before the end of Task A so a lead (not a lag) of three days should be added: Task A FS-3 Task B. Using finish-to-finish (FF) is incorrect because that will add a dependency that will require both tasks to finish at the same. Task A FS-2 Task B is incorrect because the lead is only two days.

Reference: *PMBOK® Guide,* 4th Ed., page 138

80. D. A lag refers to a desired delay in the start of a successor task relative to the predecessor task. It can be used with any dependency type, such as FS+, SF+, SS+, or FF+, to indicate the amount of the desired delay. However, if other predecessor tasks in the network will take longer to complete than the lag (delay period), the successor task will start later than anticipated.

Reference: *PMBOK® Guide,* 4th Ed., page 140

81. A. A lead refers to a desired acceleration in the start of a successor task relative to the predecessor task and can be used with any dependency type, such as FS-, SS-, or FF-, to indicate the amount of the desired acceleration.

Reference: *PMBOK® Guide,* 4th Ed., page 140

82. B. Finish-to-start is the most commonly used type of logical relationship because most successor tasks start after the completion of the predecessor tasks.

Reference: *PMBOK® Guide,* 4th Ed., page 138

83. C. Finish-to-finish means the completion of the work of the successor depends on the completion of the work of the predecessor. This is a more accurate description compared to just stating that both tasks (predecessor and successor) must finish at the same time.

Reference: *PMBOK® Guide,* 4th Ed., page 138

84. B. The precedence diagramming method uses boxes or rectangles to represent activities. All other options describe the activity-on-arrow (AOA) diagramming method.

Reference: *PMBOK® Guide,* 4th Ed., page 138

85. A. The Estimate Activity Resources process must be coordinated with the Estimate Costs process so that the resources and the corresponding costs can be estimated properly. This is particularly important if there are certain specialized skills or knowledge that the project team may not be familiar with. Estimate Activity Resources and Estimate Costs are part of the Planning process group. The other options are incorrect because they all occur as part of the Monitoring and Controlling process group.

Reference: *PMBOK® Guide,* 4th Ed., pages 141–142

86. B. Resource leveling is typically used to resolve resource allocation issues and/or to level out the peaks and valleys of resource usage. Fast tracking is incorrect because it will typically require more resources. You can negotiate for additional resources, but it is premature if the scheduling problem can still be resolved through resource leveling. It does not make sense to finalize a schedule if there are still outstanding resource issues.

Reference: *PMBOK® Guide,* 4th Ed., page 156

87. D. In Create WBS process, the primary objective of decomposition is to divide the work into smaller, manageable components (work packages) so that costs and activity durations can be reliably estimated. The other options are farfetched relative to the Create WBS process and are therefore incorrect.

Reference: *PMBOK® Guide,* 4th Ed., page 118

88. B. As with any mathematical model that relies on historical information, it is important that the parameters be easily quantifiable so that the calculations can be performed accurately. Such parameters must be scalable, too, so that the model can be used regardless of the activity duration that must be estimated. Although it helps to have an experienced model developer, it is usually more important that subject matter experts provide the correct historical data and parameters so that the model can be built accordingly.

Reference: *PMBOK® Guide,* 4th Ed., page 146

89. A. The activity list should include descriptions of each activity to ensure that the project team members will understand how the work is to be done. The other options are either incorrect or inaccurate descriptions of the Define Activities process.

Reference: *PMBOK® Guide,* 4th Ed., pages 145, 426

90. C Activity attributes provide additional information by identifying the various components of the activity. Although the details may evolve over time, they typically include activity ID, WBS ID, activity name, resource requirements, assumptions, constraints, and other pertinent information. The activity list is incorrect because it captures all of the scheduled activities in the project. The milestone list documents the key milestones (key events) in the project. Decomposition is a tool or technique, not an output.

Reference: *PMBOK® Guide,* 4th Ed., pages 134, 136

91. A. The activities in templates can also contain a list of resource skills and their required hours of effort, identification of risks, expected deliverables, and other descriptive information. However, it does not prompt or encourage creative thinking.

Reference: *PMBOK® Guide,* 4th Ed., page 135

92. C. The outputs of the Develop Schedule process are project schedule, schedule baseline, schedule data, and project document updates.

Reference: *PMBOK® Guide,* 4th Ed., page 152

93. C. The most common technique for simulation is Monte Carlo analysis, in which a distribution of probable results is defined for each activity and used to calculate a distribution of probable results for the total project. In this case, Monte Carlo analysis can be used for what-if scenario analysis during schedule network analysis (the Develop Schedule process) to account for schedule risks.

Reference: *PMBOK® Guide,* 4th Ed., page 156

94. A. Crashing is a form of schedule compression in which more resources are used to finish an activity in less time. It is feasible to also use crashing in other ways that don't involve resources, such as paying more to expedite a package. Crashing will shorten overall project duration as long as the tasks crashed are part of the critical path. Option C is also an accurate statement but option A is a *better* choice because it addresses an important factor in crashing—applying it to tasks on the critical path.

Reference: *PMBOK® Guide,* 4th Ed., page 156

95. C. Fast tracking is defined as compressing the project schedule by overlapping activities that normally would be done in sequence, such as design and construction. Crashing will shorten overall project duration as long as the tasks crashed are part of the critical path. *Accelerating*, a made-up term, and objectives management are both out of context relative to the question.

Reference: *PMBOK® Guide,* 4th Ed., pages 157, 435

96. A. A Gantt chart graphically depicts the project schedule by listing the activities on the left and dates across the top. Horizontal bars are displayed next to each activity, with the start and end dates proportionally positioned to their corresponding dates across the top. In some project management software, a Gantt chart can be filtered to just show the milestones or displayed as a project schedule network diagram. The resource loading chart is out of context relative to the question.

Reference: *PMBOK® Guide,* 4th Ed., page 436

97. D. The other options are all false because you cannot calculate a projected completion date with earned value because information is missing concerning which completed tasks are on the critical path. An SPI less than 1 does mean that the project is behind schedule, while a CPI greater than 1 indicates that the project is under budget. Although many explanations are possible, this combination often indicates that the project team is not progressing as planned in regard to schedule and that they have spent more money than budgeted relative to what they have completed.

Reference: *PMBOK® Guide,* 4th Ed., pages 181–183

98. D. All activities have baseline start and finish dates once project execution begins. Completed tasks also have an actual start date. Forecast remaining time has meaning only in the context of an activity in progress.

Reference: *PMBOK® Guide,* 4th Ed., pages 165–168

99. B. Project cost management should also consider the effect of project decisions on the cost of using the project's product. This broader view of project cost management is often called life-cycle costing. The other options are made-up terms.

Reference: *PMBOK® Guide,* 4th Ed., pages 165–168

100. C. The project management plan is technically the correct answer because it includes the cost management plan. However, because the cost management plan has a specific focus relative to cost management and control, it is a *better* answer. The cost performance baseline is used for comparison purposes with the actual results. The cost baseline captures the total project costs plus management reserves if applicable. The project funding requirements show how the funds will be released incrementally as per the cost baseline.

Reference: *PMBOK® Guide,* 4th Ed., page 181

101. A. Cost budgeting involves allocating the overall cost estimates to individual activities or work packages to establish a cost baseline for measuring project performance. Cost budgeting is tied to the Determine Budget process but it is not used as a basis for project contingency reserve.

Reference: *PMBOK® Guide,* 4th Ed., page 174

102. C. The question describes the key characteristics of a cost management plan. The project management plan is technically correct because the cost management plan is part of it. However, the cost management plan is a *better* answer because it is more specific compared to the project management plan. The risk management and quality management plans are incorrect because they focus on risk and quality, respectively. However, it is worth noting that the cost management plan covers the financial aspects of risks (budgets, contingencies, and management reserves).

Reference: *PMBOK® Guide,* 4th Ed., pages 165–166, 278

103. A. Cost estimating is an iterative process and it should be refined as more information becomes available. In the early phase of the project, ROM can be in the range of ±50% and it can be as narrow as ±10% in the later phases of the project.

Reference: *PMBOK® Guide,* 4th Ed., page 168

104. B. Information on the cost of many categories of resources often is available from project files, commercial cost-estimating databases, and project team knowledge.

Reference: *PMBOK® Guide,* 4th Ed., pages 165–166

105. A. QC is part of prevention and appraisal costs (cost of conformance). In contrast, costs of nonconformance include costs associated with expenses *after the fact* such as rework, warranty, or reputational loss due to defects.

Reference: *PMBOK® Guide,* 4th Ed., page 195

106. D. Pricing is a business decision that uses the cost estimate as only one of many considerations.

Reference: *PMBOK® Guide,* 4th Ed., pages 165, 168

107. D. Activity cost estimating, a part of the Estimate Costs process, involves developing an approximation (estimate) of the costs of the resources needed to complete project work. The other options are simply incorrect or inaccurate.

Reference: *PMBOK® Guide,* 4th Ed., page 174

108. C. The cost and accuracy of bottom-up estimating are driven by the size and complexity of the individual activity or work package: smaller activities increase both cost and accuracy of the estimating process. Bottom-up estimating entails estimating lower (smaller) components of work and then aggregating them together to arrive at the total overall estimate.

Reference: *PMBOK® Guide,* 4th Ed., pages 172, 427

109. C. Factors would also be acceptable if speaking strictly in a mathematical or statistical context, but parameters is a better answer in project management terminology.

Reference: *PMBOK® Guide,* 4th Ed., page 173

110. D. More specifically, this is an example of analogous estimating, which is one of the top-down estimating techniques. Analogous estimating entails using the parameters of previous, similar activity (for example, size, weight, quantity, and so forth) to arrive at an estimate for a comparable parameter or to measure future activity.

Reference: *PMBOK® Guide,* 4th Ed., pages 171, 427

111. D. A code of accounts reflects the financial reporting structure that has been defined in the work breakdown structure (WBS). It is a numbering system that enables the project management team to uniquely identify each component of the WBS. The other options are incorrect because they inaccurately describe the characteristics of the code of accounts.

Reference: *PMBOK® Guide,* 4th Ed., pages 121, 429

112. D. Variance analysis involves comparing actual project results to planned or expected results. Cost and schedule variances are the most frequently analyzed, but variations from the plan in areas of scope, resource, quality, and risk often are of equal or greater importance.

Reference: *PMBOK® Guide,* 4th Ed., page 127

113. A. The Create WBS process involves breaking down a project into smaller, more manageable units. The lowest level of decomposition that occurs within a WBS is known as a work package. The other options are incorrect because as their names imply, they deal with defining, verifying, and controlling the project scope, not breaking down the project into work packages.

Reference: *PMBOK® Guide,* 4th Ed., page 117

114. B. Power has the potential to influence behavior, change the course of events, overcome resistance, and get people to do things. The negative sense, of course, derives from the fact that attempts to reconcile these interests result in power struggles and organizational games that sometimes can take on a thoroughly unproductive life of their own.

Reference: *PMBOK® Guide,* 4th Ed., pages 417–421

115. C. The best available option is to deny the customer request and ask him to open a change request. Adding the feature because it will improve customer satisfaction or because otherwise the project will not be accepted by the customer are not the best answers, because adding this feature will impact your project deliverables and this feature was not in the initial scope of work. Talking with the sponsor about the benefits of having the feature assumes that you have some information about the benefits of the feature of which the sponsor is not aware, which may not be true.

Reference: *PMBOK® Guide,* 4th Ed., pages 352–356

116. A. Documentation must be prepared and distributed demonstrating that the client or sponsor has accepted the product of the project phase or major deliverable(s). Such acceptance may be conditional, especially at the end of a phase.

Reference: *PMBOK® Guide,* 4th Ed., page 99

117. C. Cost estimates are the quantitative assessments of the likely costs of the resources required to complete project activities. The other options are incorrect because the question described that specific term.

Reference: *PMBOK® Guide,* 4th Ed., page 52

118. B. Maslow's Hierarchy of Needs, a theory in psychology proposed by Abraham Maslow, are physiological, safety, social, esteem, and self-actualization, from bottom to top. It appears that the functional managers are more concerned with their employment needs (job security) instead of their physiological (general well-being), self-esteem (confidence), and social (love/belonging) needs.

Reference: *PMBOK® Guide,* 4th Ed., page 418

119. B. The project managers must know that the final project report is more than just a log of events. For that, they must be aware of all the people who receive the project report (organizational structure) and also the statutory requirements that must be met while framing such a document. Along with this knowledge, they must possess good communication skills. Therefore, knowing the organizational structure and statutory requirements is the correct answer. The other options are not relevant to this task.

Reference: *PMBOK® Guide,* 4th Ed., pages 12–15, 102

120. A. Closed procurements is an output from the Close Procurements process. Organizational process asset updates is the only other output of this process.

Reference: *PMBOK® Guide,* 4th Ed., pages 341–344

121. A. Change request status updates are an output of Perform Integrated Change Control and not a tool.

Reference: *PMBOK® Guide,* 4th Ed., page 42

122. C. Documenting the criteria in the project charter is the correct option. Project selection criteria are used in scoring models to measure relative benefits of projects. Scoring models are one type of benefit measurement methods; cash flow analysis techniques are another type. All of these benefit measurement methods are used as project selection methods. The feasibility study can be done after the project is chosen to make sure it is economically viable. Cash flow analysis is not valid for this question.

Reference: *PMBOK® Guide,* 4th Ed., page 85

123. D. Knowledge of constraints and assumptions will facilitate the process of starting the Plan Communications process. This process also produces the communications management plan.

Reference: *PMBOK® Guide,* 4th Ed., page 247

124. C. In general, differences between and among stakeholders should be resolved in favor of the customer.

Reference: *PMBOK® Guide,* 4th Ed., page 46

125. A. Benefit-cost analysis involves estimating tangible and intangible costs (outlays) and benefits (returns) of various project and product alternatives. The other options are simply incorrect relative to the question.

Reference: *PMBOK® Guide,* 4th Ed., page 52

126. D. Parametric estimating can be used for the entire project of for a part of it, and in conjunction with other estimating techniques. Using the parametric estimating technique with other estimating techniques can yield accurate results, and the results will be even more accurate if the historical information is reliable, the model is scalable, and the parameters are quantifiable.

Reference: *PMBOK® Guide,* 4th Ed., page 150

127. B. Contractual agreements with unions or other employee groups are constraints for the project. Constraints are limitations that the project team will need to work with while performing project work.

Reference: *PMBOK® Guide,* 4th Ed., page 115

128. B. The objective of procurement negotiations is to build a lasting relationship. Maximizing monetary return in itself should not be a main consideration because quality, responsibilities and authority, applicable terms and law, and technical and business management approaches also should be considered.

Reference: *PMBOK® Guide,* 4th Ed., pages 332–333

129. C. The Project Cost Management knowledge area includes the following three processes: Estimate Costs, Determine Budget, and Control Costs. The Develop Schedule process is part of the Project Time Management knowledge area, and the Perform Qualitative Risk Analysis process is part of the Project Risk Management knowledge area.

Reference: *PMBOK® Guide,* 4th Ed., page 43

130. C. Simulation involves calculating multiple project durations with different sets of activity assumptions that account for schedule risk. The most common technique is called Monte Carlo analysis, in which a distribution of probable results is defined for each activity and is used to calculate a distribution of probable results for the whole project.

Reference: *PMBOK® Guide,* 4th Ed., page 156

131. A The Close Project or Phase process confirms that the project has met all requirements of the product of the project. This process includes a formal acceptance by the customer that the project results and deliverables have met the requirements of the organization. In cases where the project ended prematurely, the Close Project or Phase process investigates and documents the reasons for early project termination.

Reference: *PMBOK® Guide,* 4th Ed., pages 99–100

132. B. Confronting, also known as problem solving, focuses on solving the problem to create a win-win resolution, while not being combative.

Reference: Reference: *PMBOK® Guide,* 4th Ed., page 240

133. C. By completing the work and submitting his status reports on time consistently, John deserves the award for that particular week. The Develop Project Team process indicates that only desirable behavior should be recognized, which makes the other options acceptable also. However, John is the only one who showed consistency which makes him the best option, whereas Janet, Larry, and Maria showed a lack of balance in their personal life and performance.

Reference: *PMBOK® Guide,* 4th Ed., page 234

134. C. Fringe Benefits = Basic Wages × 0.3 = $14.00 × 0.3 = $4.20

Overhead = (Basic Wages + Fringe Benefits) × 0.6 = (14.00 + 4.20) × 0.6 = $10.92

Loaded Cost = Basic Wages + Fringe Benefits + Overhead = 14.00 + 4.20 + 10.92 = $29.12

When estimating the costs, do not be confused by the statement that overhead costs are 60% of wages plus fringe benefits. Think of it as 60% of (wages plus fringe benefits), not (60% of wages) plus fringe benefits.

Reference: *PMBOK® Guide,* 4th Ed., page 52

135. B. The company's risk tolerance level, obtained from the enterprise environmental factors and documented in the risk management plan, is the only criterion appropriate for this situation. The other options are incorrect because they are tools and techniques.

Reference: *PMBOK® Guide,* 4th Ed., page 276

136. B. The project charter, a formal authorization of the existence of the project, is created before resource requirements can be estimated.

Reference: *PMBOK® Guide,* 4th Ed., page 442

137. A. The formula for calculating cost variance (CV) is EV – AC, and scheduled variance (SV) is EV – PV.

CV in A is $600, CV in B is –$100, and CV in C is $200. Option D requires that you calculate SV, since only the EV and PV values are provided. The SV for D is –$100. The negative variance in B is bad, meaning that the project is performing poorly in terms of costs and so apparently the project manager's attention should go to this case, but in option A, the variance is the largest (even though it's positive, which can be seen as something good). Without additional information, we would be required to select the negative cost variance, making B the correct option. But we're not able to make this determination, since we cannot properly evaluate all four options, as option D is SV, not CV. The correct answer must be A because the question is "Which of the following activities most deserves your immediate attention?"; the question is not whether the situation is good or bad. In this case, the large variance might be a failure in planning the cost of the project.

Here is another way of answering this question:

Cost Performance Index (CPI) = Earned Value (EV) / Actual Cost (AC)

Schedule Performance Index (SPI) = Earned Value (EV) / Planned Value (PV)

CPI for B is 0.97, for C is 1.07, and for A is 1.25, while SPI for C is 0.97. This means that for every dollar spent on B, C, and A, we so far got back $0.97 (not bad), $1.07 (good), and $1.25 (very good), respectively, and for A, we completed 97% (not bad) of the originally planned job. In case A, the variance is the largest. The fact that case A is under budget does not mean that the cause of the variance should not be investigated; either the execution is incomplete, or the plan was significantly off as a consequence of poor scheduling. Either case probably requires some corrective action. Hence, a project manager's immediate attention will go to A.

Reference: *PMBOK® Guide,* 4th Ed., pages 181–188

138. C. Verify Scope is a part of the Monitoring and Controlling process group.

Reference: *PMBOK® Guide,* 4th Ed., page 43

139. C. The formula to calculate the total number of potential communication channels is [n(n–1)] / 2, where *n* is the number of stakeholders. Therefore, you need to calculate all of the options to see which option will give you 300: [5(4)] / 2 = 10; [10(9)] / 2 = 45; [25(24)] / 2 = 300; and [30(29)] /2 = 435. Therefore, the answer is 25.

Reference: PMBOK® Guide, 4th Ed., page 253

140. B. The formula to calculate the total number of potential communication channels is [n(n–1)] / 2, where *n* is the number of stakeholders. Because the three projects are distinct, you can calculate them separately and then just add up the number of potential communication channels. ([5(4)] / 2) + ([7(6)] / 2) + ([9(8)] / 2) = 67.

Reference: *PMBOK® Guide,* 4th Ed., page 253

141. D. Actual cost is $90,000. It is less than target cost, so you can retain the target profit ($10,000) and also can get 30% incentive from excess profit of $10,000 (= 110,000 – 90,000 – 10,000), which is $3,000. Hence, the contractor gets Actual Cost + Target Profit + Incentive = $90,000 + $10,000 + $3,000 = $103,000 Note: This is also less than ceiling price ($120,000).

Reference: *PMBOK® Guide,* 4th Ed., page 322

142. C. As a project manager, you have a responsibility to all the stakeholders in the project. In this case, the community's interests have to also be taken into account. Therefore, informing all stakeholders is the correct answer.

Reference: *Project Management Professional (PMP®) Handbook*, pages 48–49

143. B. TCPI, a ratio of the remaining work relative to the funds remaining is calculated using the formula: work remaining (BAC – EV) / funds remaining (BAC – AC or EAC – AC) = ($1,000 – $900) / ($1,000 – $950) = 2. Estimate at completion (EAC) was not provided in the question so that variation of the formula cannot be used in the calculations.

Reference: *PMBOK® Guide,* 4th Ed., page 186

144. D. A determination cannot be made whether a project is considered complete based on isolated earned value management (EVM) values. The project or phase is formally complete after the completion of the Close Project or Phase process.

Reference: *PMBOK® Guide,* 4th Ed., pages 65, 184

145. A. Work performance measurements help produce project activity metrics by comparing actual versus planned progress. Planned and actual schedule performance, and actual cost performance are useful metrics, but they must be compared against another value (for example, plan vs. actual) for them to provide useful and actionable information. Therefore, planned vs. actual technical performance is a *better* option.

Reference: *PMBOK® Guide,* 4th Ed., page 208

146. C. The question described the tasks that are typically associated with the Plan Procurements process. It appears that Debra is working on some of the outputs of the Plan Procurements process: procurement management plan, procurement statements of work, make-or-buy decisions, procurement documents, source selection criteria, and change requests.

Reference: *PMBOK® Guide,* 4th Ed., page 313

147. D. Marketplace conditions; products, services, and results in the marketplace; suppliers; typical terms and conditions; and unique local requirements are collectively known as enterprise environmental factors (EEFs) that can influence the Plan Procurement process. Past performance or reputation is incorrect because it is just one of the many examples of EEFs that include both internal and external environmental factors.

Reference: *PMBOK® Guide,* 4th Ed., page 320

148. B. Risk response strategies for negative risks or threats include avoid, transfer, and mitigate. In contrast, risk response strategies for positive risks or opportunities consist of exploit, share, and enhance. It is worth noting that acceptance can be used for both negative and positive risks. Therefore, the correct answer is enhancement.

Reference: *PMBOK® Guide,* 4th Ed., pages 303–305

149. C. Profit/Loss Amount = Target Cost – Actual Cost = \$130,000 – \$150,000 = –\$20,000

Incentive Amount = 20% of (–\$20,000) = –\$4,000

Without any constraints, Profit Amount = Target Profit + Incentive Amount = \$15,000 – \$4,000 = \$11,000

Total Price = Actual Cost + Profit Amount = \$150,000 + \$11,000 = \$161,000

But there is a ceiling, so the modified Total Price is \$160,000.

The seller therefore earns a profit of \$10,000 = \$160,000 – \$150,000.

Reference: *PMBOK® Guide,* 4th Ed., page 322

150. B. The scope management plan is prepared during the planning sessions of the Develop Project Management Plan process. This planning effort occurs prior to the five processes (Collect Requirements, Define Scope, Create WBS, Verify Scope, and Control Scope) of the Project Scope Management knowledge area.

Reference: *PMBOK® Guide,* 4th Ed., page 103

151. B. Model A: Per month cost = \$600.

Model B: Total Cost = (4 years × \$1,000) + \$20,000 + (\$550 × 48 / 10) = \$26,640. Per month cost = \$26,640 / 48 = \$555.

Reference: *PMBOK® Guide,* 4th Ed., pages 168, 316

152. D. The question describes the risk register, which is the only output of the Identify Risks process. The Plan Risk Management process produces the risk management plan as an output. The Perform Qualitative Risk Analysis and Plan Risk Responses processes introduce updates to the risk register.

Reference: *PMBOK® Guide,* 4th Ed., pages 160, 170, and 274

153. B. The first step is to prepare a plan for risk management followed by risk identification. After that, the project team can perform risk analysis to better understand the nature of the risks. The response to this analysis may result in any of the other three actions.

Reference: *PMBOK® Guide,* 4th Ed., page 283

154. C. From a project perspective, option A is a budget question that should have nothing to do with the functional manager's area of responsibility. Option B is a schedule question, which is the responsibility of the project manager. Option D is a project justification, which should be part of the scope statement and hence the project manager's responsibility. The correct answer is option C, because in a matrix organization, the functional manager holds the resources and the project manager will have to request and negotiate for the resources needed for the project.

Reference: *PMBOK® Guide,* 4th Ed., pages 26–30

155. C. Reward power is based on the right of the project manager to offer rewards to others for doing what is wanted or expected of them and to be able to deny rewards for failing to do what is desired or expected of them. Reward is a positive recognition and penalty is a negative reward; both are derived from the same source, called reward power.

Reference: *PMBOK® Guide,* 4th Ed., pages 225, 234

156. C. A matrix organization requires the team to report to two different managers (the project manager and the functional manager). That makes team and morale building more difficult than in the other cases, where loyalty is to only one manager.

Reference: *PMBOK® Guide,* 4th Ed., pages 29–30

157. C. Compromising involves settling a dispute where two or more parties agree to accept less than what they originally wanted. Neither side fully wins or loses. Smoothing has high concern for relationships because disagreements appear minimized but has low concern for personal goals because it does not lead to a permanent solution; conflict simply is pushed to a future time. Forcing shows high concern for personal goals because one party gets exactly what it wants, but low concern for relationships because it is not a good way to foster team building. The correct answer is problem solving, which shows high concern for personal goals because the solution that works best for both parties is chosen and there is less room for ambiguity, leading to high concern for relationships, as the conflict is permanently resolved.

Reference: *PMBOK® Guide,* 4th Ed., page 240

158. D. Assembling the team and asking them to prepare an agenda for topics to discuss is not correct because the client has requested the meeting and should have their own agenda of topics to discuss. If this were a regular status review meeting, the project manager would prepare the agenda. Making sure that each team member has assignments in preparation for any handouts and needed plans is a waste of time when the true agenda is not known. Making sure that the team has increased productivity so the client can see how efficiently your team is performing is not correct, as no mention has been made of concerns about productivity. Updating the schedules and assuming an active role since the client has not given you the specifics of the potential problem is the correct response given that the client has not described the true problem. If this is a design issue that the client had originally signed off, you might need an appropriate change request. There is also a chance that the client has misunderstood the issue and there is no real problem. All the project manager can do is assume an active role and be prepared to resolve the stakeholder's issues.

Reference: *PMBOK® Guide,* 4th Ed., page 246

159. D. A milestone refers to a significant point or event in the schedule. Option A is not correct because that is part of the description for a project schedule network diagram. Option B is not correct because a milestone chart is not a bar chart. Option C is not correct because a milestone chart is not a project schedule network diagram. Option D is the correct response.

Reference: *PMBOK® Guide,* 4th Ed., page 51

160. D. Contract negotiation is not correct because it happens after a proposal is selected. Expert judgment is not correct because it does not assign numbers to qualitative data. Screening system is not correct because a screening system defines the minimum criteria and not a ranking. As part of Conduct Procurements process, a weighting system is the correct response because it is the system that is used to establish a negotiating sequence by ranking all proposals by quantifying qualitative data to minimize the effects of personal prejudice in evaluating each proposal.

Reference: *PMBOK® Guide,* 4th Ed., pages 328–329

161. A. Determine Budget is the process of aggregating the estimated costs of individual activities or work packages to establish a cost baseline. Control Costs is the process of influencing the factors that create variances, and controlling changes to the project budget. Estimate Costs is the process of developing an approximation of the costs of the resources needed to complete the project. Crashing is a type of project schedule compression technique used to decrease the total duration of the project and is not a process.

Reference: *PMBOK® Guide,* 4th Ed., page 176

162. A. Progressive elaboration refers to adding detail to the scope as more information becomes known. This is so because usually little information is known about the scope during the early phases of the project. The planning and re-planning, making changes to scope (presumably via an approved change control process), and fully documenting each step of the project are potential consequences of progressive elaboration.

Reference: *PMBOK® Guide,* 4th Ed., page 442

163. A. A feasibility study involves determining whether there is a potential solution, assessing whether the project should be carried out, selecting a solution from multiple options, and finding out if a project is feasible given available resources. If not, none of the further analysis or planning shown in the other options is appropriate. You could do a cost-benefit analysis, which is less time-consuming, but it does not show a complete picture.

Reference: *PMBOK® Guide,* 4th Ed., page 45

164. B. To get a competitive advantage in the market, to meet a customer's needs, and to meet regulatory requirements are the only common reasons to start a project, based on the options provided. Other reasons include taking advantage of an opportunity, capitalizing on technological advances, and addressing a market demand.

Reference: *PMBOK® Guide,* 4th Ed., pages 5–6

165. C. Logic indicates that the most influence/control is be exerted during the beginning (Initiating) phase of a project, when the most decisions remain to be made.

Reference: *PMBOK® Guide,* 4th Ed., pages 44–45

166. D. The project charter should be approved by someone outside of the project such as a sponsor, Project Management Office (PMO), or portfolio steering committee. An approved project charter authorizes the project manager to start the project. Except for finding an executive who is conveniently available to approve the project charter, all of the other options are valid reasons for the project charter to be signed by the appropriate executive.

Reference: *PMBOK® Guide,* 4th Ed., pages 73–74

167. A. All the options except A involve people or organizations. Archived records involve data, not the judgment of people.

Reference: *PMBOK® Guide,* 4th Ed., page 77

168. A. It is not possible to make detailed estimates of costs this early in a project. Detailed cost estimates should be expected in the Planning process group.

Reference: *PMBOK® Guide,* 4th Ed., pages 115–116

169. B. The project management team seems most logical, given that everyone would have bits and pieces of the information needed to define the project, but only the project manager has overall responsibility for this task.

Reference: *PMBOK® Guide,* 4th Ed., pages 13, 444

170. D. The sponsor authorizes resources (typically money), and the project manager then allocates the resources. This question could be confusing unless you know what makes up the project charter. The sponsor authorizes the use of resources in the charter; everyone else just uses (spends or allocates) resources.

Reference: *PMBOK® Guide,* 4th Ed., pages 74, 449

171. A. Most things within the organization (internal) are OPAs. Option A is the only option that is outside an organization and is considered an enterprise environmental factor.

Reference: *PMBOK® Guide,* 4th Ed., page 439

172. C. The correct answer is to define the project scope in order to produce the project scope statement. After defining the project scope statement, the project manager can perform other activities such as providing estimates, forming the CCB, and performing risk analysis.

Reference: *PMBOK® Guide,* 4th Ed., page 103

173. C. *Reviews*, *audits*, and *walkthroughs* are some of the terms used to mean inspections. However, depending on the organization or application area, these terms may have a specific meaning (for example, review is not treated or performed exactly the same way as audit).

Reference: *PMBOK® Guide,* 4th Ed., page 124

174. B. EEFs include both internal and external environmental factors that can impact the success of the project such as organizational culture, industry standard, technical infrastructure, existing human resources, marketplace conditions, and more. Processes and procedures is the correct answer because they are part of organizational process assets (OPAs), not an EEFs.

Reference: *PMBOK® Guide,* 4th Ed., pages 14, 32

175. B. The RAM is completed in the Planning phase as part of the Develop Human Resources process. Inputs of the Develop Project Charter process include project statement of work, business case, contract, enterprise environmental factors, and organizational process assets.

Reference: *PMBOK® Guide,* 4th Ed., pages 75–76, 220

176. C. Used to identify project and product requirements, some of the group creativity techniques are brainstorming, nominal group technique, Delphi technique, idea/mind mapping, and affinity diagram. Unanimity is not a group creativity technique but a group decision-making technique. Along with majority, plurality, and dictatorship, unanimity is one of the methods for reaching a group decision.

Reference: *PMBOK® Guide,* 4th Ed., page 108

177. B. The correct answer is decision tree analysis. As part of quantitative risk analysis, a decision tree diagram divides a decision to be made into decision nodes (for example, whether to buy or build a product). Each decision node is then further divided into chance nodes with their corresponding scenario probabilities (for example, 60% probability for a strong demand and 40% probability for a weak demand) to arrive at the net path values. *Decision rating chart* and *tree decision analysis* are made-up terms and are therefore incorrect options. The Delphi technique is a group creativity technique that relies on anonymous feedback from a team of experts.

Reference: *PMBOK® Guide,* 4th Ed., pages 299, 432

178. D. Important interpersonal skills include leadership, team building, motivation, communication, influencing, decision making, political and cultural awareness, and negotiation. Scope, cost, and time estimating techniques are not part of managing interpersonal relationships.

Reference: *PMBOK® Guide,* 4th Ed., page 417

179. A. Using a schedule network, the critical path method calculates the theoretical start (early and late) and end dates without regard to resource limitations. This method helps identify the sequence of activities that has the least amount of scheduling flexibility. Gantt charts graphically depict the project schedule by listing the activities on the left and dates across the top. Horizontal bars are displayed next to each activity with the start and end dates proportionally positioned to their corresponding dates across the top. A dependency diagram captures the relationships between predecessor and successor activities. The Delphi technique is a group creativity technique that relies on anonymous feedback from a team of experts.

Reference: *PMBOK® Guide,* 4th Ed., page 154

180. A. Resource leveling is typically used to resolve resource allocation issues and/or to level out the peaks and valleys of resource usage. Resource leveling often extends the end date of the project.

Reference: *PMBOK® Guide,* 4th Ed., page 156

181. D. The Estimate Activity Durations process uses the following tools and techniques: expert judgment, analogous estimating, parametric estimating, three-point analysis, and reserve analysis. Schedule network analysis is a technique of the Develop Schedule process.

Reference: *PMBOK® Guide,* 4th Ed., pages 131, 149–151

182. D. Resource requirements (labor, materials, equipment, services, facilities, contingency costs, and so forth) are needed to develop a detailed project cost estimate. The cost management plan covers how to plan, structure, and control costs but it will not have enough details for the development of a detailed project cost estimate. It is the same situation for the procurement management plan. The project charter formally authorizes the project and will not have enough information for cost estimation purposes.

Reference: *PMBOK® Guide,* 4th Ed., pages 77–78, 168, 296, and 324

183. A. Reduction in funding, in most cases, will likely impact the schedule, scope, or some other aspect of the project (for example, procurement, risks, and so forth). The first step is to assess how the funding reduction will impact the delivery of the project objectives. Based on the impact assessment, the project parameters such as scope can be adjusted accordingly via a formal change request process. Until an impact assessment has been completed, it is premature to assume that the project will be delayed and that the resources should be adjusted. Likewise, work should not be completed arbitrarily (for example, perform what the budget allows) because of potential impacts or risks to the remainder of the project. It is acceptable to renegotiate for adequate funding. However, based on the question, it appears that a decision was already made, perhaps outside of the control of the project.

Reference: *PMBOK® Guide,* 4th Ed., pages 175, 216

184. B. A projectized organizational structure typically experiences the most team anxiety at project closure because the team will be disbanded at the conclusion of the project. Unless they have another project lined up, they may find themselves unemployed. In functional, weak matrix, and strong matrix structures, they would typically just go back to their usual day-to-day roles after fulfilling their temporary project roles.

Reference: *PMBOK® Guide,* 4th Ed., page 31

185. C. The critical path may be reduced by adding resources (more workers to shorten the duration of a task or to enable the completion of certain tasks at the same time). Performing tasks at the same time, either in parallel or staggered, is one way of crashing (shortening) the schedule. Eliminating free floats is not relevant in reducing the critical path because it will not have an impact on the critical tasks.

Reference: *PMBOK® Guide,* 4th Ed., pages 154–155

186. B. A bottom-up estimate will most likely accurately reflect the actual cost of the project because it is based on estimating the lower-level activities, which are easier to estimate accurately. The activity estimates are then totaled to arrive at the overall project cost estimate. Analogous and top-down estimates tend to be less accurate than bottom-up estimates because they are based on larger units (for example, estimating the entire project by comparing it to a similar project or estimating the work by project phases). The project's budget is based on estimates and without knowing how it was calculated, it is difficult to determine whether it is accurate.

Reference: *PMBOK® Guide,* 4th Ed., page 172

187. A. Using the project charter, requirements documentation, and organizational process assets as inputs, the Define Scope process produces the project scope statement as an output. As part of this process, project documents, such as the stakeholder register, requirements documentation, and requirements traceability matrix, are also updated.

Reference: *PMBOK® Guide,* 4th Ed., pages 112, 115–116

188. A. Confronting, also known as problem solving, is a conflict solution technique that will most likely to lead to lasting solutions because of the give-and-take attitude and open dialogue nature of this technique. The other options are incorrect because there is an element of yielding to the demands of the other party in order to resolve the conflict, such as achieving a certain degree of satisfaction (compromising), pushing one's viewpoint at the expense of the other (forcing), and emphasizing areas of agreements versus differences (smoothing).

Reference: *PMBOK® Guide,* 4th Ed., page 240

189. A. Team-building activities can be categorized into tasks and processes. Examples of tasks include establishing goals, defining roles, and agreeing on procedures. Interpersonal behavior with focus on communication, conflict management, motivation, and leadership are examples of processes.

Reference: *PMBOK® Guide,* 4th Ed., page 418

190. A. The Estimate Activity Resources process generates the activity resource requirements by using as inputs the activity list, activity attributes, resource calendars, enterprise environmental factors, and organizational process assets. Given this, the project manager will need the activity list first and may need to reference organizational chart, RBS, and RAM in order to finalize the activity resource requirements.

Reference: *PMBOK® Guide,* 4th Ed., page 131

191. C. Quality control is monitoring specific project results to determine if they comply with relevant quality standards. Outputs of the Perform Quality Control process include quality measurements, validated changes, validated deliverables, organization process assets updates, change requests, project management plan updates, and project document updates.

Reference: *PMBOK® Guide,* 4th Ed., page 206

192. A. Regardless of the project size and complexity, the CCB is responsible for approving and rejecting change requests as part of the Perform Integrated Change Control process. However, the composition of the CCB may vary depending on the project. The idea champion, project manager, sponsor, and other stakeholders can be members of the CCB. The roles and responsibilities are defined in the configuration control and change control procedures.

Reference: *PMBOK® Guide,* 4th Ed., pages 94, 428

193. B. Similar to bar charts, a milestone chart is produced as part of the Develop Schedule process and it identifies only the scheduled beginning or ending of major activities or deliverables. The milestone chart would be suitable for sending monthly updates to the CEO. The other options are incorrect because they will have too many details that the senior managers and executives may not care about.

Reference: *PMBOK® Guide,* 4th Ed., page 157

194. B. There are six general techniques for resolving conflicts: withdrawing/avoiding, smoothing/accommodating, compromising, forcing, collaborating, and confronting/problem solving. Directing and controlling are not valid conflict resolution methods.

Reference: *PMBOK® Guide,* 4th Ed., page 240

195. B. All the options will lead to quality improvement. However, given that it is better to prevent problems from occurring in the first place instead of catching the defects later in the process through inspection, increasing the number of inspections will contribute the *least* to quality improvement compared to ISO 9000 certification and continuous process improvement. PDCA is the basis for continuous improvement.

Reference: *PMBOK® Guide,* 4th Ed., pages 191, 213

196. B. As part of the Plan Quality process, the COQ divides the costs into conformance and nonconformance. Cost of conformance includes preventive costs and appraisal costs whereas the cost of nonconformance consists of internal and external failure costs. Relative to the question, the options are not related to COQ except for ensuring conformance to requirements.

Reference: *PMBOK® Guide,* 4th Ed., page 195

197. A. After adding resources to shorten the duration of the tasks in the critical path, it is imperative that the project manager calculate the new critical path to determine the implications to risks (for example, new risks may arise or existing risks may increase as a result of the resource adjustments) and other project components. After identifying the critical path, further analysis can be performed on the lag, duration, and resources on the critical and near-critical activities.

Reference: *PMBOK® Guide,* 4th Ed., page 154

198. B. Activity duration estimates indicate the number of work periods required to complete a scheduled activity. The other options are incorrect descriptions of activity duration estimates.

Reference: *PMBOK® Guide,* 4th Ed., page 151

199. C. Motivating the project team is one of the many interpersonal skills that a project manager must possess. Given this, it is imperative that project managers be familiar with various motivational theories, one of which is Maslow's hierarchy of needs, which includes, from bottom to top, physiological, safety, social, esteem, and self-actualization.

Reference: *PMBOK® Guide,* 4th Ed., page 418

200. D. Key project deliverables should be reviewed by executive management when completed. If a key project deliverable can take several months to complete, the executive management may ask for interim updates instead of just reviewing it upon completion. Daily and weekly executive management review would be too frequent to be effective, and ad hoc (upon request) request may be seen as too informal.

Reference: *PMBOK® Guide,* 4th Ed., page 89

201. B. As part of the Control Schedule process, the schedule variance (SV) or schedule performance index (SPI) can be used to determine whether the project is ahead, behind, or on schedule. The formula for SV is = EV – PV = 375 – 400 = –25. Negative SV indicates behind schedule. Use the formula SPI = EV / PV = 375 / 400 = 0.9375. An SPI less than 1 indicates the project is behind schedule.

Reference: *PMBOK® Guide,* 4th Ed., pages 162, 182

202. A. The cost performance index (CPI) measures the value of completed work relative to the completed work. CPI = EV / AC = 375 / 325 = 1.15. A CPI greater than 1 is good. The cost variance (CV) can also be used to assess the project costs. CV = EV – AC = 375 – 325 = 50.

Reference: *PMBOK® Guide,* 4th Ed., pages 182–183

203. D. The Pareto chart rank-orders the problems from high to low so the correct answer is 350, 250, 200, and 180 (B, E, C, and D).

Reference: *PMBOK® Guide,* 4th Ed., pages 201–211

204. B. As described in the question, the project manager used several interpersonal skills such as leadership (marching people toward a common goal), communication (written and verbal), influencing (getting people to cooperate toward a common goal), decision making (consultation and consensus), and negotiation (win-win results). However, the main gist of the question points to the political and cultural awareness of the project manager, and she leveraged the other interpersonal skills to resolve the issues.

Reference: *PMBOK® Guide,* 4th Ed., pages 417, 419–421

205. A. Inputs to the Close Project or Phase process are the project management plan, accepted deliverables, and organizational process assets.

Reference: *PMBOK® Guide,* 4th Ed., page 65

Index

B

C

D

E

F

G

H

I

N

O

P

Q

R

S

T

U

V

W

Z